First Edition, second printing.

ISBN: 978-1-7394874-8-5

Programming ESP32

Learn MicroPython Coding and Electronics

Simon Monk

To my brother Andrew Monk. An inspiring and generous human being.

Preface

Learning to program, like any skill, requires practice and simple straight-forward examples that you can work through. In this book, you will be led step-by-step through example programs, intermingled with explanations and background.

The first chapter is an introduction to the ESP32 and the various development boards that use this chip.

Chapter 2 will help you get your computer set up and ready to start programming ESP32. This includes installing USB drivers and Thonny – a good code editor to get started with.

In Chapter 3, you will start interacting with Python and learn how to write simple programs using conditional statements and loops. You will also learn how to control the built-in LED on your ESP32 board.

Chapters 4 to 6 explore Python further, developing a complex Morse code translation program that blinks out messages using the built-in LED. Along the way you will learn about Python lists, dictionaries, functions and modules.

Chapters 7 to 9 look at how you can use your ESP32 to interact with other electronics, including LEDs, servomotors and sensors of various types.

Chapter 10 shows you how your ESP32 can connect to your local network and hence out to the world-wide web. You will learn how to make your ESP32 act as a web server and also interact with web services on the internet.

Chapter 11 shows how you can attach OLED display modules and Neopixel displays to your ESP32 using a clock example.

The final chapter is concerned with more advanced features of the ESP32 including advanced input/output features such as high-speed PWM, timers, interrupts and sound generation.

Code Download

All the code examples used in the book are available for download from github, here:

```
https://github.com/simonmonk/prog_esp32
```

You can download all the files as a ZIP archive (see Page 32).

The book's website

For errata, and further information on the book, see the book's web page at: https://simonmonk.org/esp32

Electronics Hardware

You can learn Python Programming and carry out some simple electronics, such as turning the the built-in LED on and off, using just your ESP32 board. However, at some point, you will want some basic components and tools to learn about the electronics side of using an ESP32.

You can find out a lot more about useful parts in Chapter 8, but if you have some of the following, it will give you a good start:

- Solderless breadboard – makes it easy to wire together circuits without having to use a soldering iron

- LEDs – indicator lights

- Resistors – particularly useful values are 1kΩ and 470Ω

- Push switches

- Jumper wires

You will find good value starter kits of electronic parts from various Chinese suppliers on online marketplaces. MonkMakes has a kit, originally designed for the Raspberry Pi, that contains a breadboard and many of the components used in this book. You can find more details of this here: https://monkmakes.com/pi_box_1.

This book (as you might expect from the title) is mostly concerned with programming and avoids the need for dangerous or expensive tools. A digital multi-meter (DMM) is useful but not essential – a budget one for a few dollars is fine. You don't really need a soldering iron to follow along with this book, unless you buy an ESP32 board that does not have header pins attached.

Acknowledgements

Many thanks to Ian Huntley and Mike Bassett for taking the time to provide a technical review. Your help is very much appreciated.

The breadboard layouts for this book were created using the excellent Fritzing Software (fritzing.org). Other diagrams were drawn using Inkscape (inkscape.org).

Thank you also to the creators of Thonny for making such a great beginners' code editor for Python.

Thanks, as always, to Linda for her love and support.

Contents

Contents

CHAPTER 1

Introduction

In this chapter we will look at what exactly an ESP32 board is, and introduce some of the ESP32 boards available, helping you to pick a good board to get started with.

ESP32

ESP32 is a family of processors developed by Espressif Systems (that's where *ESP* comes from in the name) that, notably, includes WiFi and Bluetooth capabilities. The processor chip itself is used on many different development boards. Sometimes the processor is placed on the board directly but, at other times, a *module* is placed on the board. This module, that encloses the processor and a chip, is covered by a metal cap; its design complies with international standards – CE marked for the EU and FCC for the United States.

Development boards that use the ESP32 come in a wide range of shapes and sizes. Figure 1.1 shows a selection of the most popular ESP32 boards.

The boards shown in Figure 1.1 are of Chinese manufacture and are readily available on international online retail markets such as eBay and Amazon. Although there are ESP32 boards available from US and European manufacturers, these are more expensive than their Chinese counterparts.

As well as the processor itself, ESP32 development boards will also include a number of other components:

- USB interface chip. This allows programs to be transferred from your computer to the board and also for the board to do things like send back messages or sensor data to your computer and with the USB interface also provides power to the board.

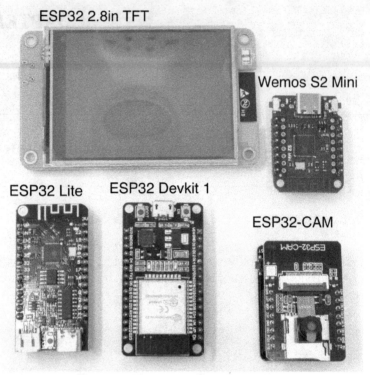

Figure 1.1. A Smorgasboard of ESP32s.

- Voltage regulator. This allows you to power the board from higher voltages than the 3.3V that the ESP32 processor itself operates on. The regulator reduces the 5V supplied by USB to 3.3V when the board is connected to USB.

- *Reset* and *Boot* buttons. These are used during setup of the processor and if the board needs reseting

- GPIO pins. These are used to connect extra electronics like LEDs, switches and sensors

- Battery Charging Chip. Some boards include a LiPo battery charging chip and a battery connector, making it easy to power your project using a battery when it's not connected to USB

Choosing an ESP32 Development Board

There are a lot of ESP32 boards to choose from. Different boards all have their pros and cons.

For the purposes of this book, we will standardize on a couple of the cheapest and most readily available boards (the ESP32 Lite and the ESP32 Devkit 1. These boards are shown alongside various other popular boards in Figure 1.1. If you have one of these boards use that but, if you are buying a new board, I would recommend the ESP32 Lite, as it fits better on solderless breadboard (see Chapter 8) and has a useful battery connector and charging circuit.

These boards are readily available online. If you don't mind waiting a while for delivery, the lowest price is obtained by ordering directly from China's Aliexpress online market. Search for *ESP32 Board* and then carefully check the description and photograph. The description should say that the board has WiFi and Bluetooth, and the photograph should look like the board in Figure 1.1. You will often get a choice of USB socket. The newest USB-C is likely to be most convenient, but any type is fine, as long as you have a lead of that type for connecting it to your computer.

It's also a good idea to look for a board that already has pin headers soldered onto it. Some boards come with the header pins that you then have to then solder on yourself.

Figure 1.1 has some other interesting boards. The ESP32 DevKit 1 is very popular, but is a bit wider than the ESP32 Lite and does not fit as well on a solderless breadboard. The Wemos S2 Mini is interesting because it uses a newer generation of the ESP32 chip that has built-in WiFi and so does not need a separate chip on the board. The board has a double row of pins, to save space, but this also makes it difficult to use with solderless breadboards.

The ESP32 2.8in TFT board – sometimes called a CYD (Cheap Yellow Display) – is interesting because most of the front of the board is taken up with a touch screen display. Figure 1.2 shows the back of the board. Here you will find various useful connectors as well as an SD card slot for extra memory for things like image files to display on the touch screen.

The ESP32-CAM board is actually a stack of two boards, the bottom board holds an ESP32, and the top board a camera capable of recording video and stills. It's the sort of thing that could be used to make a smart doorbell camera.

The ESP32 Lite and ESP32 DevKit 1

What are probably the two most popular ESP32 boards are shown side-by-side in Figure 1.3. Later in the book, when we start to attach external electronics to the boards, we will show wiring diagrams for both of these boards.

The main differences between these two boards is that the DevKit 1 uses an ESP32-WROOM module (the metal box containing the ESP32 chip itself)

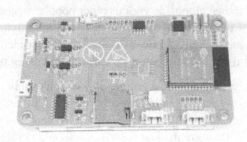

Figure 1.2. The back of the ESP32 2.8inch TFT board.

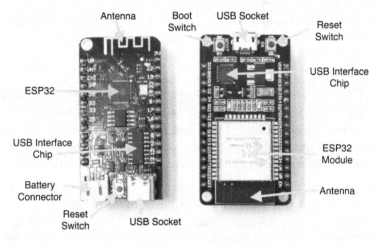

Figure 1.3. ESP32 Lite and ESP32 DevKit 1.

whereas the ESP32 Lite has the processor chip attached directly to the circuit board. As I mentioned earlier, the ESP32 Lite also has a LiPo battery charging capability and a battery connector.

General Purpose Input Output Pins

If you look closely at The two boards of Figure 1.3 you can see that the rows of pins are not labelled in the same. The two boards have mostly the same connections, but they are in slightly different places.

We will revisit these pins in detail when we come to start attaching things to them in Chapter 8.

Most of the pins are what are called GPIO (General Purpose Input Output) pins. They are *general purpose* in the sense that the particular purpose they serve is determined by the program you upload to the ESP32. They can all be configured as: *digital output* – for example turning an LED on, or as a *digital input* determining whether a push switch has been pressed.

In addition, most of the pins can be configured to be an *analog input* – measuring a voltage, perhaps from a temperature sensor. Similarly, most but not all of the pins can be used as a PWM (Pulse Width Modulation) output – for example to control the brightness of an LED.

Some of the GPIO pins also have special interface capabilities, such as various types of interface bus (SPI, I2C and UART). This allows peripherals that use those standards to communicate with the ESP32.

Appendix A shows the *pinouts* of the ESP32 Lite and ESP32 DevKit 1 boards, including such extra functions.

Programming

Programming, or *coding* if you prefer, is the process of writing a program that tells a computer, or in this case the ESP32's microcontroller, what to do. You can think of a program as a list of instructions to be carried out. For example, to make an LED blink on and off, the instructions might (in English) be as follows. If it helps, imagine someone being the processor and standing by a light switch with one hand on the switch and the other holding a watch — your job is to provide instructions that are clear and unambiguous.

1. Turn the LED on.

2. Wait for half a second.

3. Turn the LED off.

4. Wait for half a second.

5. Repeat from step 1

Following these instructions, you can see that, as well as simply performing commands like turning the LED on and off, we also need control commands like step 5 that allows earlier commands to be repeated.

When we write a program for the ESP32, we type the text for the program into an editor using a programming language. In this case, the programming language is a version of the Python programming language called *MicroPython*.

Having written the program, we save it as a file. If we then put this file onto the ESP32, then the ESP32 will run it for us.

Summary

In Chapter 2 we will install the software that you need on both the ESP32 and on your computer. You will then get started with a bit of coding.

CHAPTER 2

Getting Started

One of the great things about ESP32 boards is that all you need to get started is a USB lead to connect it to your computer and an editor in which to write your Python code before transferring it to the ESP32.

Most ESP32 boards even have a built-in LED that you can control from your program code. You can make this blink on and off, or control the brightness. It's not much, but we can add more exciting electronics as you progress through the book.

In this chapter you will learn how to install MicroPython on your board, install the software you need on your computer, how to interact over the USB connection with the ESP32, and how to transfer a program on to it. We choose to use the Thonny editor, as it has good support for the ESP32.

This chapter should leave you with an ESP32 board connected to your computer with the necessary software on your computer, and your ESP32. The steps involved in this are:

- Install USB drivers on your computer (may not be necessary)
- Install the Thonny Python editor on your computer
- Install MicroPython on your ESP32 board

USB Drivers

ESP32 boards need a USB interface so that you can connect to them with your computer to upload files onto them. Depending on your computer's operating system, you may need to download driver software for the USB interface chip used by your ESP32 board.

Drivers and operating systems get updated all the time, so it's probable that the instructions below will become out of date. If they don't work, then search for an up-to-date tutorial.

Installing the USB drivers can be problematic, but fortunately, once you have installed them on your computer, you shouldn't have to do it again.

Pick the section below depending on which board you have and also your computer's operating system (Windows or Mac OS) – most Linux distributions have the drivers pre-installed.

CHM340 USB Chip (ESP32 Lite)

If you are using the ESP32 Lite then your board will probably use the CHM340 USB interface chip. Linux will usually include a driver for this chip but, if you are using Windows or versions of Mac OS older than 10.14, then you will probably need to install driver software.

Installing the CH340 Driver on Windows

You will find the driver for Windows here: `https://tinyurl.com/5ry9287z`. Click on the option *Windows V3.5* or whatever the latest version number is. This will start a download of the file *CH341SER_WIN_3.5.ZIP*.

Open the ZIP archive and extract it, then open *SETUP*. This will start the installer running (see Figure 2.1). Windows will probably ask you to approve the installation.

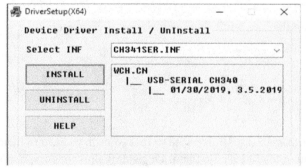

Figure 2.1. The Windows installer for CH340.

Installing the CH340 Driver on MacOS

If you have MacOS 10.14 (Mojave) or more recent, then there is nothing to do – the driver should be pre-installed. You can check your mac version by clicking on the Apple logo in the top left of the screen and selecting *About This Mac*.

To install the driver on versions of MacOS prior to 10.14, first download it from here: `https://tinyurl.com/5ry9287z`.

Apple try to discourage its users from downloading programs from the Internet, so you effectively need to opt-in to getting non App Store apps by opening *Security & Privacy* in *System Preferences* before you try and run the installer for the driver. In the Preferences panel select *App Store and identified developers*.

Note that you will probably need to click on the padlock icon and enter your Mac password before you can make this change to your preferences.

The driver download is a ZIP file called something like *CH341SER_MAC_1.5.ZIP* (your download will probably have a higher version). When opened, this will unzip to a folder containing a package called *CH34x_Install_V1.5.pkg*. Open the installer and follow the instructions.

CP2102 USB Chip (ESP32 DevKit 1)

Aside from the CH340, the other much-used USB interface chip is the CP2102 used by the ESP32 DevKit 1 amongst others. This is preinstalled in most Linux installations. Mac and Windows users will need to download a driver.

Installing the CP2102 Driver on Windows

Download the driver installer from the CP2102 manufacturer's website here: `https://tinyurl.com/3r4fsvzh` and select the latest windows driver (*CP210X Windows Driver*).

This will download a ZIP archive called *CP210x_Windows_Driver.zip*. Extract the archive file and then run *CP210xVCPInstaller_x64* to start the installer (Figure 2.2).

Figure 2.2. The Windows installer for CP2102.

Installing the CP2102 Driver on MacOS

Download the driver installer from the CP2102 manufacturer's website here: `https://tinyurl.com/3r4fsvzh` and select the latest Mac OS download.

The driver page talks about VCP. If you were wondering what this meant, a VCP is a Virtual Com Port. That is, a driver that allows the USB interface chip to act like a serial interface to your computer.

Once you have downloaded the archive file and unzipped it, you will see a file called something like *Mac_OSX_VCP_Driver.zip*. Once unzipped, a folder of the same name will be created containing the driver disk image called *SiLabsUSBDriverDisk.dmg*, Open this to mount the disk (Figure 2.3)

Open *Install CP210x VCP Driver.app* to start the installer and then follow the prompts (Figure 2.4).

Figure 2.3. The CP210x Mac Disk Image.

Figure 2.4. The CP210x Mac Installer.

Identifying the USB chip

You might have noticed, when you bought your ESP32 board, that the USB chip was specified - either CH340 or CP2102. If you didn't, or you are not sure, you can tell by looking at the board.

The black rectangular chip nearest the USB connector on the ESP32 board is generally the USB interface chip. Fortunately the CH340 and CP2102 come in quite different packages. The CH340 is rectangular with pins down two sides (Figure 2.5), whereas the CP2102 (Figure 2.6) is square with pins on all four side that are just tucked under the body of the chip. If you set a lamp at a shallow angle and photograph the chip with your phone and then zoom right in, you should see tiny writing on the top of the chip that identifies it.

Note that some newer versions of the ESP32, that use the ESP32-S2 variant, do not need a separate USB interface chip – there is built-in support for USB. What's more, these chips do not need a driver to be installed on Windows, Mac or Linux. So if your board does not seem to have any USB interface chip at

Figure 2.5. Closeup of a CH340 USB Chip.

Figure 2.6. Closeup of a CP2102 USB Chip.

all, then this might be the reason. The Wemos S2 Mini shown in Figure 1.1 is one such board.

Installing the Thonny Editor

Now that our computer has the necessary drivers to be able to communicate with an ESP32 board, we need to download software that will allow us to write code and transfer it to the ESP32.

To install programs onto your ESP32, you will need a second computer. This can be a Windows, Mac or Linux computer, or even a Raspberry Pi such as the Raspberry Pi 5 or 400. The Thonny Python editor will do this job, and is available for all the popular computer platforms - see https://thonny.org.

Windows

Click on the Windows link in the downloads section of the Thonny homepage. This will download an exe installer. Once it has downloaded, run the installer (Figure 2.7). You can accept all the defaults that the installer offers, clicking Next to take you from step to step.

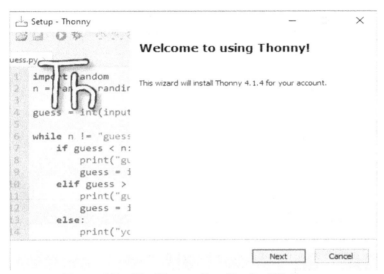

Figure 2.7. The Thonny Installer for Windows.

When the installer has finished, click on the *Finish* button and you should now see Thonny on your Start menu.

Mac OS

On Mac OS, Thonny is installed using a package installer. Click on the Mac download link and, when the .pkg files has finished downloading, open it (Figure 2.8).

Click on *Continue* after each step of the installation until installation is complete.

Linux

Installing Thonny on a Linux distribution is best done using the pip3 Python Package Installer by running the command:

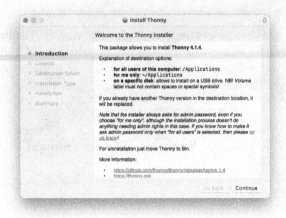

Figure 2.8. The Thonny Installer for Mac OS.

```
$ pip3 install thonny
```

If your distribution uses the *apt* package manager, you can install using the following command.

```
$ sudo apt install thonny
```

Installing Python on your ESP32

There are many programming languages available for the ESP32 and, while some ESP32 boards are shipped with MicroPython pre-installed, this is by no means guaranteed. It's also quite likely that, even if your board does have MicroPython on it, it will have an old version. So it's a good idea to follow the instructions here to install the latest version of MicroPython onto the board.

The easiest way to install MicroPython on the board is to use the built-in feature of Thonny that allows MicroPython to be installed onto your ESP32 board directly from Thonny. This feature of Thonny only became available in version 4.1 of Thonny so, if you happen to have an old version of Thonny on your computer, you might like to upgrade, to take advantage of this helpful new feature.

Whichever method you are using, you will have to put your ESP32 board into *download* mode before you can install new firmware (MicroPython) on it. The procedure for this varies between boards. It's easiest with boards, like the ESP32 Devkit 1, that have both BOOT and EN (reset) buttons. Here, see Figure 2.9, if the BOOT button is depressed when the board is powered up or reset by pressing the EN button, then the ESP32 will enter upload mode and wait for your computer to send it new firmware. This should not be confused with uploading a Python program onto the board. Essentially you are installing the Python programming language on the ESP32.

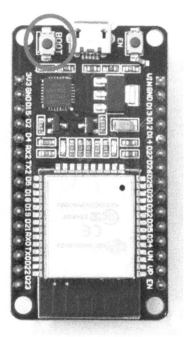

Figure 2.9. The ESP32 DevKit 1 BOOT button.

So, to put your ESP32 DevKit 1 style of board into upload mode, you do the following:

1. Connect the board by USB to power it.

2. Press and keep the BOOT button depressed.

3. Momentarily press the EN button to reset the board.

4. You can now release the BOOT button and your board will be in upload mode.

You don't get any confirmation that the board is in upload mode, you won't find out if this worked until you try and upload firmware onto it, as described later. But once in upload mode, it will remain in upload mode until it is reset or loses power.

When you press the BOOT button on an ESP32 DevKit 1, the GPIO pin 0 is connected to GND (ground). Our other recommended board (the ESP32 Lite) does not have a BOOT button, so you have to use a wire to connect GPIO 0 to GND to put it into upload mode. You can do this using female to female jumper wire connected directly between the two pins as show in Figure 2.10 or you can plug your board into solderless breadboard (Figure 2.11) and use a male-to-male jumper wire.

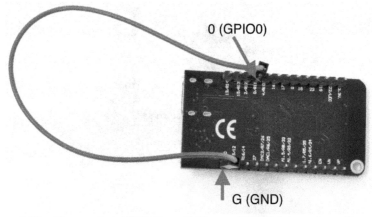

Figure 2.10. Connecting GPIO 0 to GND on a ESP32 Lite using a Female to Female Jumper Wire.

Jumper wires are short wires with either sockets (female) or plugs (male) on each end of the wire. They are available as all three combinations of connector: plugs on both ends (male to male) sockets on both ends (female to female) or one of each (female to male). They are a handy way of making connections between ESP32 board pins and peripherals, or to other pins on the ESP32 board.

As we start to attach extra electronics to our ESP32 in Chapter 8 we will be using a breadboard, so now might be a good time to plug your ESP32 board into breadboard as shown from below in Figure 2.11.

If you are like me, then once you have attached the ESP32 to the breadboard, you probably won't need to take it out again.

The *Project Box 1 for Raspberry Pi* from MonkMakes (`https:// monkmakes.com/pi_box_1`) includes a breadboard, jumper wires and a selection of components to get you started.

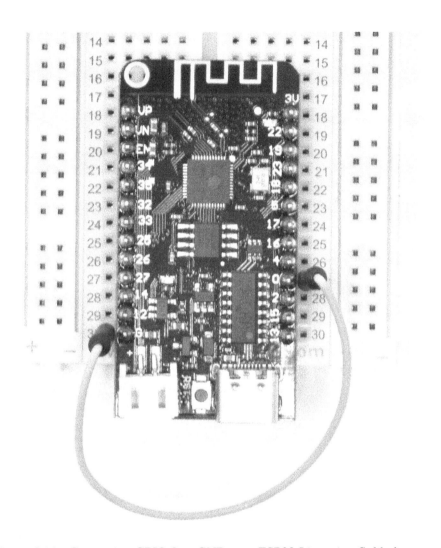

Figure 2.11. Connecting GPIO 0 to GND on a ESP32 Lite using Solderless Breadboard.

Using Thonny to install MicroPython

Now is the time to find out if your USB driver installation was successful. So, put your ESP32 board into Upload mode (see Page 15).

The feature to install MicroPython onto a board is on the Thonny *Options* Panel. On Windows and Linux, you will find this on the *Tools* menu, called just *Options...* On a Mac, you will find this as *Preferences...* from the *Thonny* menu. Whatever you do to open it, the panel will look something like Figure

2.12.

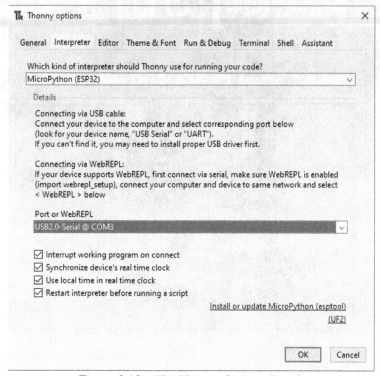

Figure 2.12. The Thonny Options Panel.

On the Options panel, click on the *Interpreter* tab and select *MicroPython (ESP32)* from the interpreter drop-down. If your USB driver is working then, when you click on the *Port or WebREPL* dropdown, you should see an option like the one shown in Figure 2.13 (*USB2.0-Serial @ COM3*).

When you switched the Interpreter to *MicroPython (ESP32)* an extra option appeared at the bottom of the options panel called *Install or update MicroPython (esptool)*. Click on this option and a new panel like Figure 2.13 appears.

Select *ESP32* from the MicroPython family dropdown. On the *variant* dropdown select *Espressif - ESP32 / WROOM* which should work for most ESP32 boards.

Finally click on the *Install* button. You will see progress messages appear as Python is installed onto the ESP32.

Once complete, Python is installed and ready for action.

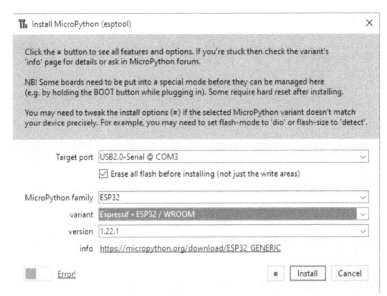

Figure 2.13. The Thonny Options Panel - Installing MicroPython.

Thonny

Before we get going with Thonny, let's first check that we are all set up to use an ESP32 board. So, connect your board to your computer, start Thonny, and then open the *Options* pane and select the *Interpreter* tab (see Figure 2.12).

Confirm that the *interpreter* drop-down is set to *MicroPython (ESP32)* and that *Port or WebREPL* is set to a com port as shown in Figure 2.12. If you are using a Mac or Linux computer the port name will be different, and contain the words *tty* and *usb*.

If you don't have a serial port shown in Thonny, then please see Appendix B for possible reasons and solutions.

Close the *options* panel and return to the main Thonny window. There are three main areas to the Thonny editor, the toolbar at the top, the editor area in the middle and the area labeled *Shell* at the bottom.

The area of Thonny labelled *Shell* allows you to run individual Python commands directly on the ESP32, without having to write a whole program. This is a great way to experiment with Python and to try things out before committing them to a program file.

The Shell area should look like it does in Figure 2.14. The characters >>> are the standard Python prompt, asking you to enter some Python code to be run. In this case, we can see that 2 + 2 has already been run and Python has given us the result (4).

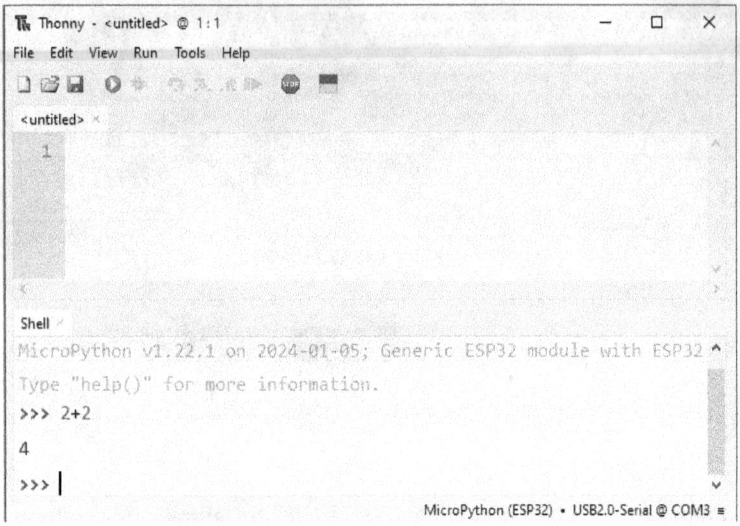

Figure 2.14. The Thonny Shell.

Let's take a moment to think about what is going on here. The Python programming language is running on the ESP32 and we are using Thonny to interact with the ESP32 board. It is the ESP32 board that is adding 2 and 2 together, not your computer. Your computer is just displaying the result using Thonny.

In Thonny, programs are contained in text files, and the toolbar area provides access to buttons that allow you to create new program files, as well as loading and saving files from your file system. Python programs all have the extension of *.py*.

As well as buttons for managing the program files, you will also find buttons in the toolbar for running the program being edited, for stopping the running program, and also for using the debugger that allows you to diagnose problems with your program by stepping through it one step at a time.

The editor area allows you to modify the program code. You will notice that there are line numbers in grey at the side of the editor. This is useful as, when there are problems running the code, the description of the problem will refer to a line number, helping you to find the source of the problem.

Summary

Setting up your computer and *flashing* MicroPython onto your ESP32 board can be difficult and does not always go smoothly. If you have problems, then work your way through this chapter again, and you may also find some help in Appendix B (Throubleshooting).

Now that our computer and ESP32 board are set-up and ready for action, we can start getting to grips with MicroPython.

CHAPTER 3

MicroPython Basics

In this chapter, we will explore some of the basics of Python using our ESP32 board, getting used to some key ideas like numbers, variables and ways of making our code take different paths. We will also use the ESP32 board's built-in LED – as a first step towards a project, that we will gradually develop, to use the LED to flash out Morse Code messages.

The Thonny Shell

Thonny's interactive Shell is a great way of experimenting with and learning about MicroPython. Let's try a little experiment to turn the built-in LED on and off. Type the following in the Shell.

```
>>> from machine import Signal, Pin
```

Later on we will learn about `Signal`' and `Pin` - for now, all we need to know is that they are external libraries that we can use to avoid having to write all the code ourselves.

After the `import` commands, type one of the two commands below. If you have an ESP32 Lite then type:

```
>>> led = Signal(22, Pin.OUT, invert=True)
```

If on the other hand you have a ESP32 Devkit 1, then type:

```
>>> led = Signal(2, Pin.OUT)
```

Finally type:

```
>>> led.on()
```

When you type this last line, the built-in LED should light. Type the following line to turn the LED off again.

```
>>> led.off()
```

The reason for using `invert=True` with the ESP32 Lite, is that the LED logic is inverted on that board so, setting the pin high turns the built-in LED off rather than on. Whenever we use the built-in LED on an ESP32 Lite, we use `invert=True`.

Numbers

Let's carry out a few more experiments using Thonny's Shell feature. Start Thonny, and connect your ESP32 board to your computer. Start by repeating an experiment from Chapter 2 and type 2 + 2 after the >>> prompt.

Here we are adding the two integers 2 and 2 together, to get the integer 4.

There are two important types of numbers in Python: *integers* (or *ints* for short) which are whole numbers like 1, 2, 3 etc. and floating point (*floats*) that have a decimal place such as 1.1, 2.5, 10.5 etc. In many situations, you can use them interchangeably. Now try typing the following after the prompt:

```
10 + 5.5
```

As you can see, the result is 15.5 as you would expect where we are adding an integer to a float. As you can see, the result is the float 15.5. This is an example of a general rule when you mix integers and floats.

Try out a few sums for yourself. If you want to do multiplication use * and for division use /.

Note that, whenever you use the Shell, things will be proceeded by >>> and the ESP32 board's response will start on a new line.

Enter the following into the Shell.

```
>>> from random import randint
>>> randint(1, 6)
5
```

You are likely to get a different answer from 5. Repeat the `randint(1, 6)` line a few times. Rather than typing in the command every time, you can use the up-arrow key to recall your last command - very useful. You will see a number between 1 and 6 - you have made a dice, of sorts!

The first line you typed imports the `randint` function from the module `random`.

Again, random is just a module containing all sorts of useful bits of Python – including the `randint` function, which provides random integers.

Modules are used to contain Python code that you may not need all the time. This helps to keep the size of the code small enough to run on a ESP32 board.

Variables

Try typing the following line into the Shell. This will assign the value 10 to the variable x

```
>>> x = 10
```

You can put spaces either side of the = sign or not, it's a matter of personal preference. I think it looks neater with spaces, so that's the standard I will be sticking to in this book.

This line of code is using a variable called x. Now that the variable x has been given (or assigned) the value 10, try just typing x in the Shell

```
>>> x
10
```

Python is telling us that it remembers that the value of x is 10. You don't have to use single letter names for variables. for variables - you can use any word that starts with a letter, and you can and you can include numbers and the underscore (_) character.

Sometimes you need a variable name that's made up of more than one human language word. For example, you might want to use the name `my number` rather than x. You're not allowed to use spaces in variable names, so a common trick is to use the underscore character – to give `my_number`. Another way to do this is to use what is called camel case (think humps) - giving `myNumber`.

By convention, variables usually start with a lowercase letter. If you see variables that start with an uppercase letter, it usually means that they are what are called *constants*. That is, they are variables whose value is not expected to change during the running of the program, but that you might want to change before you upload your program onto your ESP32 board.

Returning to our experiments in the Shell, now that Python knows about x you can use it in sums. Try out the examples below:

```
>>> x + 10
20
>>> x = x + 1
>>> x
11
```

In this last command (x = x + 1) we have added 1 to x (making 11) and then assigned the result to x, so that x is now 11. Increasing a variable by a certain amount is such a common operation that there is a shorthand way of doing it.

Using += combines the addition and assigning a value into one operation. Try the following:

```
>>> x += 10
>>> x
21
```

You can also use round brackets (and) to group together parts of an arithmetic expression. For example, when converting a temperature from degrees Centigrade to degrees Fahrenheit, multiply the temperature in degrees C by 9/5 and then add 32. Assuming that the variable c contains a temperature in degrees C you could write the following in Python:

```
>>> c = 20
>>> f = (c * 9/5) + 32
>>> f
68.0
```

The (and) are not strictly necessary here, because MicroPython will automatically perform multiplication and division before it does addition, but it can make the code easier to follow by using brackets.

Strings

Computers are really good at numbers, that is after all what they were originally created for. However, as well as doing math, computers often need to be able to use text. Most often, this is to be able to display messages that we can read.

In computer-speak bits of text are called *strings*, and you can give a variable a string value (just like we did for numbers). Type the following into the Shell.

```
>>> s = 'Hello'
>>> s
'Hello'
```

So first we assign a value of *Hello* to the variable s. The quotation marks around *Hello* tell Python that this is a string and not the name of a variable.

You can use either single or double quotation marks, but they must match.

We then check that s does contain *Hello* by typing s on the Shell.

Rather like adding numbers, you can also join strings together (called concatenation).

Try the following in the Shell:

```
>>> s1 = 'Hello'
>>> s2 = 'World'
>>> s = s1 + s2
>>> s
'HelloWorld'
```

This might not be quite what we were expecting. Our strings have been joined together, but it would be better with a space between the words. Lets fix this:

```
>>> s = s1 + ' ' + s2
>>> s
'Hello World'
```

This adds three strings together, the middle string containing just a single space.

Programs

The commands that we have been typing into the Shell are single-line commands that just do one thing. You can see how typing them one after the other leads to them being run or executed one after the other. For example, revisiting this example:

```
>>> s1 = 'Hello'
>>> s2 = 'World'
>>> s = s1 + s2
```

```
>>> s
```

The four command lines are executed one after the other as fast as we can type them. If you put all four lines into a file, and then tell the ESP32 board to run the whole file, you will have written a program.

Let's try that now – but first we need to make one slight change. When you use the Shell and simply type the name of a variable, the Shell will display the value in that variable. However, when running the same commands as a program, you must use the print function for any value that you want to display.

So, click on the *New* icon in Thonny's toolbar to create a new program, and then type the following lines into the editor window (Figure 3.1).

```
s1 = "Hello"
s2 = "World"
s = s1 + s2
print(s)
```

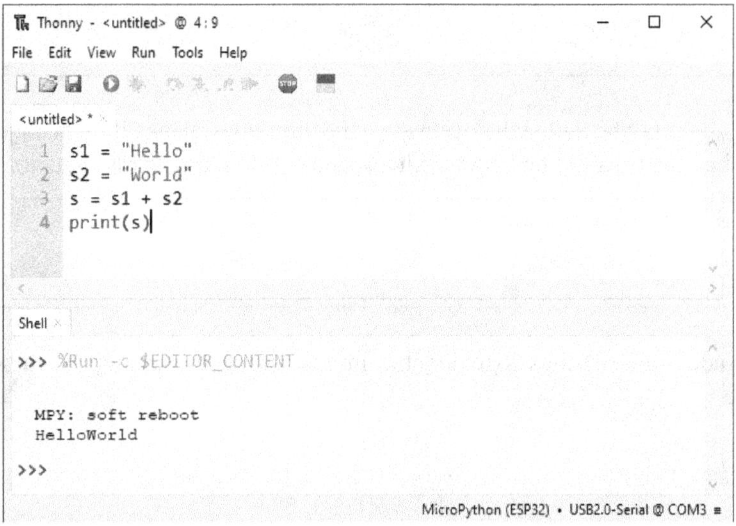

Figure 3.1. Hello World in Thonny.

Notice how the program is labelled <untitled> at the top, because we have not yet saved the file anywhere. At the moment, our tiny four-line program exists only in Thonny – we should save it as a file, so that it won't be lost when we exit Thonny. To do this, click on the *Save* button (looks like a floppy disk) or chose *Save* from the *File* menu. This will open the popup shown in (Figure 3.2).

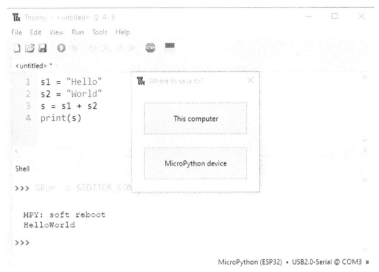

Figure 3.2. Saving to your computer vs saving to the ESP32 board.

Choosing the *Save to your computer* option will open a file dialog window, where you will select a location on your computer to save the file. If you select *Save to MicroPython device*, then the file will be saved to the ESP32 board's built-in file system. The choice as to where to save the program file has some important implications. Saving to the ESP32 board itself has the benefit that, once saved to the ESP32 board, the ESP32 board no longer needs Thonny or your computer to be able to run the program. Also, if you name the file *main.py*, it will automatically run that program every time it is powered up and will be retained by the ESP32 even when not powered.

Saving on the ESP32 board does run the risk that, should something happen to your ESP32 board, your program could be lost. So, when we are still experimenting with small programs, it's fine to save it on the ESP32 board but, if you were writing a longer program, that you wanted to keep safe, you would save it on your computer first and then (when ready to deploy it properly onto the ESP32 board) you would save a copy of the program onto the ESP32 board.

For now, let's save it on the ESP32 board – so select the option *MicroPython device* and then give the file the name *main.py* as shown on Figure 3.3.

Now that the file has been saved, click on the *Run* button and you will see the message *helloworld* appear in the Shell – the Shell is now the area where program output appears.

Figure 3.3. Saving a program onto the ESP32 board.

Looping Forever and Blinking

Generally speaking, you don't want the program running on a ESP32 board to end – it's not like running an app on your computer, that you can just quit when you have finished using it. A program on a ESP32 board will basically run until you unplug it. This is why you find something called a *while loop* at the end of most ESP32 board programs.

As it's a loop in the code that continues forever, let's call it *the eternal loop* from now on.

Type the following program into the editor area of Thonny, replacing the code that was there.

```
from machine import Signal, Pin
from time import sleep

led = Signal(22, Pin.OUT, invert=True)

while True:
    led.on()
    sleep(0.5)
    led.off()
    sleep(0.5)
```

Before we look at what this code does, or actually run it, an important point about indentation needs to be made. You will notice that after the `while` `True:` line, the remaining lines are all indented by four spaces. This indicates that those lines belong inside the `while` command. Python insists that they line up exactly (or you will get an error message) and the convention is to use four spaces of indentation. Helpfully, in Thonny, when you press the TAB key you get four spaces inserted for you.

This program is going to make the board's built-in LED blink so, before running the program, we need to set the pin number to the pin number associated with your board. If you are using an ESP32 Lite, then 22 is the right number and you don't have to change anything. Table 3.1 shows the pin number for the built-in LED for a few common ESP32 boards. Use this to change 22 to the pin number for your board if necessary. Also, if you have a board that does not have inverted logic for the LED, then you don't need `invert=True` in the LED definition.

Table 3.1.: LED Pins by Board.

Board	Built-in LED Pin	Logic
ESP32 Lite	22	value(0) - Inverted
ESP32 Devkit 1	2	value(1) - Normal
ESP32-S2 Mini	26	value(1) - Normal

As before, save the program under the name *main.py* and you should see the built-in LED start to blink.

Just to prove that the code is indeed now installed on the ESP32 board, unplug the ESP32 board, quit Thonny, and then plug your ESP32 board back in. Now, the only thing that your computer is doing for the ESP32 board is supplying it with power over USB. Your ESP32 board should automatically run the program and start blinking.

Here's how the code works. The first two lines import `Pin`, `Signal` and `sleep` into our program so that we can use them. This is necessary because much of MicroPython, especially the parts specific to the ESP32 board, are kept in code modules. You will learn more about this in Chapter 6.

The variable `led` is then associated with the built-in LED by pin number.

The command `while` `True:` tells the ESP32 board to repeat all the indented lines (after the : on the end of the line) forever. The indented lines turns the LED on, insert a delays of 0.5 of a second, turn the LED off, and insert another 0.5s delay - followed by the whole cycle repeating indefinitely.

Example Code

It's a good idea to type a few programs into Thonny while you are learning. However, all the code examples for this project are available for download from GitHub at:

`https://github.com/simonmonk/prog_esp32`

so it's a good idea to download them, especially as the code examples start to get a bit longer.

To download the files, visit the link above in your browser and then click on the *Code* button and then *Download ZIP* (Figure 3.4), When the ZIP archive has downloaded, extract its contents to get the Python files contained in a folder called *prog_esp32-main*.

Figure 3.4. Downloading the Example Code from Github.

Within that folder you will find two more folders named *esp32_devkit_1* and *esp32_lite*. The programs within these two folders are duplicates of each other, except that the code uses the correct built-in LED pin number for that board, and in the case of the ESP32 Lite, flips the on/off logic for the built-in LED.

The blink program that you just wrote can be found in the code download, the file is called *03_01_blink.py*. Each program file is prefixed by the chapter number (in this case 03). Try loading this file in Thonny by clicking on the *Open* button and navigating to wherever you extracted the ZIP file to.

Blinking SOS

Morse Code is an ancient type of signalling code, that uses a system of short and long pulses, either of sound (long and short beeps) or flashes of light. We are going to use the ESP32 board's built-in LED to signal for us. In Morse code, short pulses (of sound or light) are called *dots* and long pluses are called *dashes*.

The starting point for our Morse code adventure is to blink out the emergency signal SOS (repeated) which comprises three short pulses (dots) for the letter S, followed by three long pulses (dashes) for the letter O, followed by three dots for the second S, and then repeating. Strictly speaking, the SOS should be like a word, with a gap before the next sequence of . . . ---. . ., but the convention for SOS in an emergency setting is that it is just an unbroken repetition of the letters S and O: . . . ---. . . ---. . . ---. . . .

We will define a dot as a light signal for 0.2s, a dash as a light signal for 0.6 second, and the gap between dot/dash as 0.6 second.

We can modify our blink program, so that it generates this SOS signal, by adding some more `led.on()`, `led.off()` and `sleep` commands to what is already in the eternal loop.

```python
from machine import Pin, Signal
from time import import sleep

led = Signal(22, Pin.OUT, invert=True)

while True:
    # S
    led.on()
    sleep(0.2)
    led.off()
    sleep(0.2)
    led.on()
    sleep(0.2)
    led.off()
    sleep(0.2)
    led.on()
    sleep(0.2)
    led.off()
    sleep(0.6)
    # O
    led.on()
    sleep(0.6)
    led.off()
    sleep(0.6)
```

```
led.on()
sleep(0.6)
led.off()
sleep(0.6)
led.on()
sleep(0.6)
led.off()
sleep(0.6)
```

Try running the program *03_02_sos.py*. Remember to change 22 to your board's built-in LED number if you are not using an ESP32 Lite board. You should see the ever repeating pattern of three dots followed by three dashes. Note the use of the # comment mark to tell us that the code that follows is for the letter S or O. Any text after #, on that line, is not program code, but rather comments that help the programmer and others looking at the code understand what is going on.

The code is now starting to get a little long and repetitive. In the next section we will use a *for loop* to improve this a little and, in the next chapter, we will streamline this code even more by use of something called *functions*.

For Loops

In this section you will learn about *for loops*. This means telling Python to do things a number of times rather than just once. This is similar to the eternal *while loop*, except that this is not eternal – we will specify before it starts how many times it should repeat before the program carries on and does something else (or, if there is nothing else to do, just quits).

To get a feel for how this works, type the following code into a new editor window in Thonny, or load and run the example *03_03_for_loop.py*.

```
for x in range(1, 10):
    print(x)
```

When you run the program, you should see the following output in the Shell. You may need to scroll up to see the start of the output.

```
1
2
3
4
5
6
```

```
7
8
9
>>>
```

This program has printed out the numbers between 1 and 9 rather than 1 and 10 as you might be expecting. The range command has an exclusive end point – that it, it doesn't include the last number in the range, but it does include the first.

There is some punctuation, here, that needs a little explaining. The round brackets are used to contain what are called *parameters* . In this case, range has two parameters: the start of the range (1) and the end (10), separated by a comma.

The for..in command has two parts. After the word for there must be a variable name. This variable will be assigned a new value each time around the loop. So, the first time it will be 1, the next time 2, and so on. After the word in, Python expects to see a list of things. In this case, that list of things is a sequence of the numbers between 1 and 9.

The print command also takes an argument which tells it what to display in the Python Shell. Each time around the loop, the next value of x will be printed out.

If we wanted to, we could use a *for loop* to change our SOS blinking program to repeat the dot and dash blinking. You will find this example in *03_04_sos_for.py*. It should work exactly the same as *03_02_sos.py*, but the code is a bit shorter.

```
from machine import Pin, Signal
from time import sleep

led = Signal(22, Pin.OUT, invert=True)

while True:
    # This loop does 3 dots (S)
    for x in range(1, 4):
        led.on()
        sleep(0.2)
        led.off()
        sleep(0.2)
    sleep(0.4) # Gap between S and O
    # This loop does 3 dashes (O)
    for x in range(1, 4):
        led.on()
```

```
        sleep(0.6)
        led.off()
        sleep(0.6)
    sleep(0.4) # Gap between O and S
```

The first thing to note here is that even through we want three blinks, the range has to be 1 to 4, rather than 1 to 3 because, as we mentioned before, the end point of the range is exclusive. It's easiest just to remember to set the end of the range to be one more than the number of times you want something to happen.

The other point of interest in the code is that an extra sleep of 0.4 is needed between the dots flashing and the dashes flashing, because this extra delay can no longer be built in.

if and else

Most programs are more than just a simple list of commands to be run over and over again. They can make decisions, do some commands only if certain circumstances apply. They do this using a special Python word called if and its optional counterpart else.

Let's illustrate the use of the if command by making the program ask for you to type a number into the Shell, and then say something about that number.

Type the following code into a new editor tab, or load the program *03_05_if_simple.py*.

```
response = input("Enter a number: ")
number = int(response)
if number > 10:
    print("That's a big number")
```

Run the program, noting that you must ensure the cursor is within the Shell area before typing a response. Enter a number greater than 10 – you will get a response in the Shell saying that the number is big. Run the program again, and this time enter a number less than or equal to 10 – you should not see a message this time.

Note the use of the input command, that provides a prompt in the Shell for you to enter a number, and whatever you enter will be assigned to the variable response. Whatever you type, even if its a number, will actually be a string and so must be converted to an integer using the int command before we can make comparisons with other numbers (in this case, the number 10).

The remainder of the line after the word `if` is called a *condition*. That means the condition that must be true for the code inside the `if` (indented) to run.

Conditions can use

- `<` (less than)
- `>` (greater than)
- `<=` (less than or equal to)
- `>=` (greater than or equal to).

You can also use `==` for exactly equal to and `!=` for not equal to. These comparisons can also be combined into more complex statements of logic using the words `and` and `or`.

We can add an `else` command to this code (*03_06_if_else.py*).

```python
response = input('Enter a number: ')
number = float(response)
if number > 10:
    print("That's a big number")
else:
    print("That's a small number")
```

The function `float` is used to convert the string into a floating point number.

Notice that when we get to the `else` part of the code, we remove the indentation, back to the same level as the corresponding `if`. The `print` command to be run whenever the `if` condition is NOT true must be indented.

You can see the result of running this program in Figure 3.5.

Summary

This chapter has covered quite a lot of ground in terms of basic programming concepts. If you are new to programming it can take time to make sense of these ideas, so you may like to play with the code examples so far—alter them, run them, and see what happens. Experimenting is a great way to learn. If you mess up the programs, it doesn't matter you can just download them again.

In the next chapter you will learn all about one of the most important features of the Python language – functions.

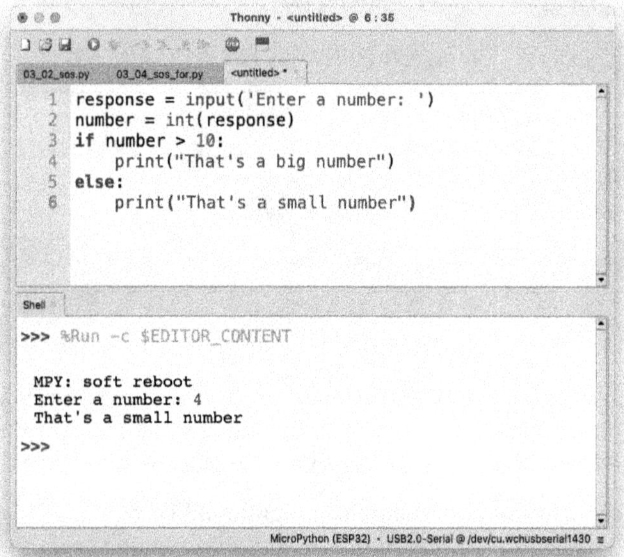

Figure 3.5.   Running the if..else example in Thonny.

CHAPTER 4

Functions

Programs have a way of running away from you as they start to get more complex. Even with the simple S.O.S. example in Chapter 3, you could see how the code was getting a bit repetitive and lengthy. Functions are one of the key tools that the programmer can use to make their code concise and easy to read.

You have already used some functions that are part of Python, such as `sleep`, `input` and `print`. However, you can and should also define your own functions when you are writing programs.

What are Functions?

A function is a little like a program within a program. When you make your own function, you give it a name and some code that should be run whenever you call the function should be run. Although programs and functions are often described as *running* when they are doing something, you will also hear the phrases *executing*, *calling* or *invoking* a function. They all mean the same thing.

A good MicroPython program will have most of its code contained in functions, with just a few lines at the bottom (the eternal loop) where all the functions you defined earlier get used.

The code contained in a function can call other functions, and should be named in a way that reflects what it does. So function names usually describe some action, such as `blink` or `throw_dice`.

Parameters

The built-in function `sleep` needs to know how long it needs to sleep for, and so takes a *parameter* in round brackets – for instance `sleep(0.5)` which tells it to sleep for half a second.

Similarly, the function `print` expects a parameter telling it what should be printed. For example, `print('Pass the salt')` tells it to print the string inside the quotes.

Let's create our own version of `print` called `print_polite`. This function will simply add the string `'please'` to the end of any message printed. You can find this example in *04_01_print_polite.py*.

```
def print_polite(text):
    print(text + ' please')

print_polite('Pass the salt')
```

Run the program and you should see the text *Pass the salt please* in the Shell. The word `def` (short for *define*) marks the start of the function definition. An important point about functions is that just because they are defined, they are not necessarily used. Just because, we have defined a function called `print_polite`, does not mean that any of the lines of code within it will be run – until we call it using the code `print_polite('Pass the salt')`.

After the word `def`, marking the start of the function definition, we have the function's name. This name must be unique within your program, and is how you will refer to the function when you come to *call* it. So, its name can't be the same as any variables, or other words that Python uses like `print`, `if` etc. It follows the same naming convention as variables, so separate multiple words with an underscore character, as we have with `print_polite`. You also need to pick a name for your function that describes what it does, so that any reader of your program has a good idea what's going on.

Let's pick up our S.O.S. example from the program *03_04_sos_for.py*.

```
from machine import Pin, Signal
from time import sleep

led = Signal(22, Pin.OUT, invert=True)

while True:
    # This loop does 3 dots (S)
    for x in range(1, 4):
        led.on()
```

```
    sleep(0.2)
    led.off()
    sleep(0.2)
sleep(0.4) # Gap between S and O
# This loop does 3 dashes (O)
for x in range(1, 4):
    led.on()
    sleep(0.6)
    led.off()
    sleep(0.6)
```

Remember that you may need to change the pin number from 22 to whatever pin is connected to the built-in LED on your board, if you are not using an ESP32 Lite.

We have already improved this over our original S.O.S. example by using `for` loops to repeat some of the blinking. We could make this code a lot neater if there were a function called `blink`, that had two parameters – the number of times to blink, and the delay between the LED changing from on to off and vice-versa.

At this point it is worth emphasising, that there is no single right way of solving a programming problem. Although I have suggested this `blink` function, it is largely to illustrate a point, and other programmers might well find better ways of doing this. Generally *better* means concise, but still easy to understand.

If we had such a `blink` function, then our eternal loop would just become:

```
while True:
    blink(3, 0.2) # blink 3 times fast (S)
    sleep(0.4)
    blink(3, 0.6) # blink 3 times slow (O)
    sleep(0.4)
```

Of course this won't work yet, because we need to define the function blink. This will look something like this:

```
def blink(times, delay):
    for x in range(1, times+1):
        led.on()
        sleep(delay)
        led.off()
        sleep(delay)
```

Notice how we have two parameters, `times` and `delay`, separated by a comma inside the parentheses. After this, there is a : to indicate the start of the function's code.

The first line of code in the `blink` function is a `for` loop that is going to control how many times the LED blinks. The `times` parameter is used here in the `range` command, with 1 added to it because the second parameter in range is exclusive.

Inside the `for` loop are the four lines of code that turn the LED on and off with a delay. Here, the fixed number (0.2 or 0.6) that we used in the old code is replaced by the parameter `delay`.

Here is the full program (*04_02_sos_function.py*)

```
from machine import Pin, Signal
from time import sleep

led = Signal(22, Pin.OUT, invert=True)

def blink(times, delay):
    for x in range(1, times+1):
        led.on()
        sleep(delay)
        led.off()
        sleep(delay)

while True:
    blink(3, 0.2) # blink 3 times fast (S)
    sleep(0.4)
    blink(3, 0.6) # blink 3 times slow (O)
    sleep(0.4)
```

Notice that we have defined the `blink` function after the `imports` and the `led` variable definition. This is where you are expected to define any functions that the program uses. We have also left blank lines between the various blocks of code – another convention that makes it easier for a newcomer to read your program.

Return Values

In the example above, the function does something – it makes an LED blink. When calling the function `blink`, we don't expect any value to come back from the function. In fact, Computer Scientists would say that `blink` is actually a *procedure* rather than a function for this very reason.

Python functions can return a value, which you can use later in the program. That is, when you call the function, it returns a value that you can assign to a variable that you can use in some way. To illustrate this, we could change the `blink` function so that, when it's finished blinking, it returns the total duration that is spent doing blinking (calculated as the number of blinks, times the duration of each blink). Here's what it would look like:

```python
def blink(times, delay):
    for x in range(1, times+1):
        led.on()
        sleep(delay)
        led.off()
        sleep(delay)
    return times * delay * 2
```

Notice the new line on the end of the function. It starts with the `return` word, followed by a sum where the number of `times` blinked is multiplied by the `delay` and then again by 2 (there are two delays). We have now specified that our function will return a number – however, we are under no obligation to use that return value in our code. So, having added that line to the function, the program *04_02_sos_function.py* would still work just fine with this version of the `blink` function. However, it does mean that we can use the return value if we want to. For example, we could change the eternal loop to look like this:

```python
while True:
    print(blink(3, 0.2))
    sleep(0.4)
    print(blink(3, 0.6))
    sleep(0.4)
```

Now, the result of each call to blink will be printed in the Shell like this:

```
1.2
3.6
1.2
3.6
1.2
3.6
1.2
```

You can find this modified program in *04_03_sos_return.py*. This addition to the `blink` function isn't actually very useful, so lets look at another example of using return values, and introduce another useful Python module.

Random Numbers

Computers behave in a very predictable manner. If you want to make them appear unpredictable, there is a Python module called random specifically for this purpose. Load the program *04_04_dice.py*. Now that you have met a few Python programs, take a good look at it and try and predict what it will do when you come to run it.

```
from machine import Pin, Signal
from time import sleep
from random import randint

led = Signal(22, Pin.OUT, invert=True)

def throw_dice():
    return randint(1, 6)

def blink(times, delay):
    for x in range(1, times+1):
        led.on()
        sleep(delay)
        led.off()
        sleep(delay)

while True:
    # Wait for Enter to be pressed
    input('Press Enter in Shell to Throw Dice')
    # Enter has been pressed
    dice_throw = throw_dice()
    print(dice_throw)
    blink(dice_throw, 0.2)
```

The areas of particular interest are mostly at the end of the program. The first new thing is that we are importing the randint function from the random module.

We have copied the useful blink function from our earlier programs, and have also defined a new function called throw_dice. The throw_dice function just returns the result of calling the built-in function randint from the random module, with parameters of 1 and 6, which is the range of random numbers that we want. You may think that defining a function that only has one line of code is barely worth-while, and you'd have a point. However we will expand this function later and, in fact, wrapping up the action of throwing the virtual dice in a function has potential advantages, if for some reason, we decided to use a twelve-sided dice, we would only have to change the code in the function rather than everywhere else.

The eternal loop starts with an `input` command. This has the effect that the program pauses until you press the Enter key in the Shell. Having pressed Enter, the function `throw_dice` is called and the result (a random number between 1 and 6) will be assigned to the variable `dice_throw`. This number is then printed and blinked.

Try pressing Enter a few times – your ESP32 board should blink out the result of the dice throw, and the number should appear in the Shell. Wait until it finishes blinking before hitting Enter again.

Named Parameters

In the previous discussion, we said that we might want to change the function `throw_dice`. Let's change it now, so that we can simulate throwing more than one dice. It would be a nice touch if the number of dice to throw were an optional parameter so that, by default, just one dice would be thrown.

Note that throwing two dice does not result in a total that is between 1 and 12, because the lowest throw you could have is a pair of 1s – be careful.

We could calculate the throw as a random number between the number of dice and six times the number of dice, but let's make the code behave as if we actually throw three dice and add up the total.

Here is the revised `throw_dice` function:

```
def throw_dice(num_dice=1):
    total = 0
    for x in range(1, num_dice+1):
        total += randint(1, 6)
    return total
```

The first thing to notice is the parameter to `throw_dice` now has a named parameter called `num_dice`. This has an `=1` after it, which means that the default value for the parameter is 1. In other words, if you do not supply a parameter (when calling the function) the function will assume that you just want to throw one dice.

We have already met named parameters, when using the built-in LED, and have used `invert=True` to invert the LED on/off logic. In this case `invert` is a named parameter with a default value of `False`.

On a stylistic note, parameters like this are the only time I prefer to use = without a space on each side.

The first line of the function creates a new variable called total, which will be used to keep the total throw.

To throw two dice, we can change the code where we call throw_dice to look like this:

```
dice_throw = throw_dice(num_dice=2)
```

We could also just write:

```
dice_throw = throw_dice(2)
```

but I think it's clearer actually to name the parameter when calling the function. You can find the full program for this in the file *04_05_dice_many.py*.

Summary

Now that we know how functions work, and how we can use them to make our programs more concise, we can start to develop some more complex programs.

In the next chapter, we will continue with the Morse code example, looking at how we can use *lists* and *dictionaries* ultimately to make ourselves a Morse code translator that, when given a sentence of text, will flash it out as dots and dashes.

CHAPTER 5

Lists and Dictionaries

In this chapter, we will work towards creating a Morse code translation program, that will convert text that we type in the Shell into Morse code flashes of light. Along the way, we will learn how to use Python lists and dictionaries.

One way to think of programs is as models of the real world that you can then experiment with. So, if we want to create a Morse code translation program, we need to have a way of modelling Morse code, and then make the translation program work using this model.

To model the Morse code, we need to be able to look up the sequence of dots and dashes that form a given letter of a message. Python has a great way of representing such things in the form of dictionaries. However, before we move on to look at Python dictionaries, let's take a look at Python lists.

Lists

So far, all of our variables have been single values: just one number or one string. Most computer languages (including Python) have a way of representing a list of values. Sometimes they are called *arrays*, *vectors* or *collections*, but in Python they are just called *lists*. Figure 5.1 shows a pictorial representation of a list, where box 0 contains the number 10, box 1 the number 34, and so on.

To create this list of numbers in Python, you could write the following code, remembering that many objects in Python are numbered from 0, not 1:

```
list_of_numbers = [10, 34, 5, 92, 16]
```

Now, type this into the Python Shell and we can try some experiments on it.

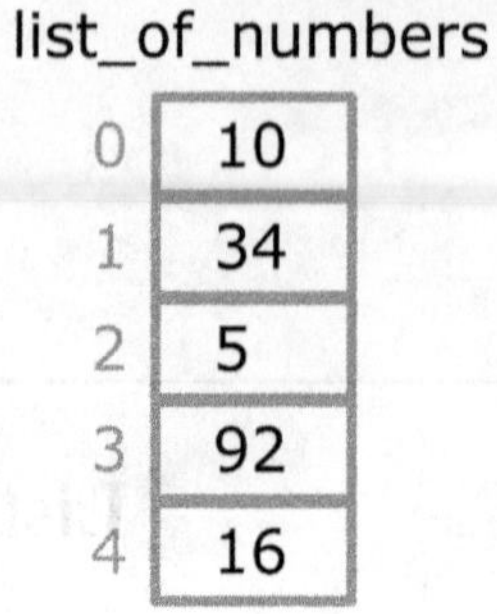

list_of_numbers

0	10
1	34
2	5
3	92
4	16

Figure 5.1. A List of Numbers.

```
>>> list_of_numbers = [10, 34, 5, 92, 16]
>>> list_of_numbers
[10, 34, 5, 92, 16]
>>>
```

When you type the variable name `list_of_numbers` for a second time, Python will display the numbers within square brackets.

To access individual elements from a Python list, you need to specify the position (also called *index*) of the element of the list that you want. Positions start at 0, so the first element of this list is at position 0 and you can retrieve it like this:

```
>>> list_of_numbers[0]
10
>>>
```

If you wanted the second element of the list, you would type:

```
>>> list_of_numbers[1]
34
>>>
```

You can also change the value at a particular index position in the list like this:

```
>>> list_of_numbers[0] = 1
>>> list_of_numbers
[1, 34, 5, 92, 16]
>>>
```

If you want to manipulate your list in other ways, such as sorting it or reversing its order, you can do things like this:

```
>>> list_of_numbers.sort()
>>> list_of_numbers
[1, 5, 16, 34, 92]
>>> list_of_numbers.reverse()
>>> list_of_numbers
[92, 34, 16, 5, 1]
>>>
```

You can also add new elements to the list as shown below.

```
>>> list_of_numbers = [10, 34, 5, 92, 16]
>>> list_of_numbers.append(12)
>>> list_of_numbers
[10, 34, 5, 92, 16, 12]
>>>
```

If you don't want the new element at the end of the list, then you can use insert which takes two parameters: the first being the position to insert the new element at, and the second is the value to insert.

```
>>> list_of_numbers.insert(1, 12)
>>> list_of_numbers
[10, 12, 34, 5, 92, 16]
>>>
```

Note that insert(1, 12) will insert the number 12 as the second element – the numbering always starts from 0.

Here the extra bits of code that carry out extra operations – .sort, .reverse, .append, .insert - are called methods, and we will learn more about methods when we study object-orientation later.

A List Example

There are lots more things you can do with lists but, for now, let's just say that lists are very flexible and move onto a more useful example. Later in this chapter, we will use a different type of data structure (dictionary) to make a general-purpose Morse Code translator but, in the meantime, let's look at how we could use a list of time delays to blink out S.O.S. (yet again). You can find the program in *05_01_sos_list.py*.

```
from machine import Pin, Signal
from time import sleep

led = Signal(22, Pin.OUT, invert=True)
delays_list = [0.2, 0.2, 0.2, 0.6, 0.6, 0.6]

while True:
    for delay in delays_list:
        led.on()
        sleep(delay)
        led.off()
        sleep(delay)
```

Notice a difference in the `for` loop from how we used it previously. Now the `for` loop is used to cycle through the various elements in the list, using each one in turn, rather than merely increasing a count by one each cycle.

Strings as Lists

You can treat a string as a list of individual characters. In fact, you can use the same square bracket notation to fetch individual characters from a string. Try the following in the Shell.

```
>>> s = 'ESP32 Lite'
>>> s[0]
'E'
>>>
```

You can find the length of the string by doing:

```
>>> len(s)
10
>>>
```

showing that spaces are treated just the same way as any other characters in the string.

As well as being able to access individual letters from the string, you can also select a range of letters in the string – a *substring*. For example, if we wanted the first three letters we could do:

```
>>> s = 'ESP Lite'
>>> s[0:3]
```

The number before the : is the start index of the substring and the second number is one more than the end index.

If you put a negative number in after the :, instead of counting forwards, the code counts from the end of the string, like this:

```
>>> s = 'ESP Lite'
>>> s[0:-1]
'ESP Lit'
>>>
```

One difference between strings and lists is that you cannot change a string once it has been created. So, the example below will cause an error.

```
>>> s = 'ESP Lite'
>>> s[0] = 'r'
Traceback (most recent call last):
File "<stdin>", line 1, in <module>
TypeError: 'str' object doesn't support item assignment
>>>
```

Dictionaries

Whereas, in a list, you access things by their position in the list, or you take each element of the list in turn and do something with it, a dictionary is a little different.

A dictionary holds a collection of values, but each value has a *key* associated with it that you use to access that *value*. This is rather like a regular directory – the key is the word you are interested in, and the value is the definition of that word.

For this reason, dictionaries are often called *lookup tables*. The key is usually a string, but it doesn't have to be; it could be a number or any other type of data.

Let's define a dictionary with two entries in it. It's probably a good idea to follow along with these examples in the Python Shell.

```
>>> d = {'x' : 23, 'y' : 77}
>>> d
{'y': 77, 'x': 23}
```

```
>>>
```

The first thing to notice is that, whereas when you define a list you use square brackets, dictionaries are defined using curly-braces. We have defined a dictionary with two entries in it. Each entry is separated by a comma (just like lists) but, in a dictionary, each entry must have two parts separated by a : (the key and the value). So, the first entry has as key of the string 'x' with a value 23. The second entry in the dictionary has a key of 'y' and a value of 77. Interestingly when we ask Python to display the value of our dictionary d, it does not show the two entries in the same order as when we created the dictionary. This is an important difference between dictionaries and lists. You cannot rely on the order of the entries in a dictionary staying the same. Figure 5.2 shows a graphical representation of this dictionary.

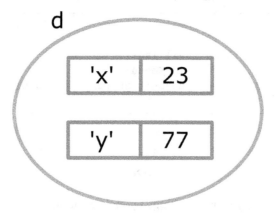

Figure 5.2. A Python Dictionary.

If we want to access a particular entry in the dictionary, we have a couple of choices. If we are certain that the key for the entry that we are going to ask for definitely exists, then we can use a square bracket notation like this:

```
>>> d['x']
23
>>>
```

However, if instead of asking for 'x' or 'y' (which exist), we asked for 'z' (which doesn't) we will get an error like this:

```
>>> d['z']
Traceback (most recent call last):
File "<stdin>", line 1, in <module>
```

```
KeyError: z
>>>
```

So, for situations where the key might not be there, but we don't want to cause an error, we can use get' like this:

```
>>> d
{'y': 77, 'x': 23}
>>> d.get('x')
23
>>> d.get('z')
>>>
```

We haven't got an error, even though they key doesn't exists, but instead nothing came back. In fact, what came back was Python's special value of None.

We can see this using print:

```
>>> print(d.get('z'))
None
>>>
```

To add an new entry to the dictionary, or indeed to replace an entry with a particular key, you can just do this:

```
>>> d['z'] = 1234
>>> d
{'y': 77, 'x': 23, 'z': 1234}
>>>
```

A Morse Code Translator

Now that we know how to use dictionaries, we can create a dictionary to represent the Morse code — then, for any letter, we can look up the sequence of dots and dashes to be flashed. You can find a table detailing the Morse code on its Wikipedia page (https://en.wikipedia.org/wiki/Morse_code).

Let's start by creating a dictionary for the letters A to Z.

```
codes = {
    'a' : '.-', 'b' : '-...', 'c' : '-.-.',
    'd' : '-..', 'e' : '.', 'f' : '..-.',
    'g' : '--.', 'h' : '....', 'i' : '..',
```

```
      'j' : '.---', 'k' : '-.-', 'l' : '.-..',
      'm' : '--', 'n' : '-.',   'o' : '---',
      'p' : '.--.', 'q' : '--.-', 'r' : '.-.',
      's' : '...', 't' : '-',    'u' : '..-',
      'v' : '...-', 'w' : '.--', 'x' : '-..-',
      'y' : '-.--', 'z' : '--..'
}
```

The positioning of the curly braces and the layout are just for ease of reading.

You can find this dictionary in the file *05_02_morse_dict.py*. The program won't do anything except assign the dictionary to the variable codes. So run it, and then try looking up various letters using the Shell like this:

```
>>> codes['a']
'.-'
>>>
```

This is telling us that the code for the letter 'a' is .- (dot dash). We cpuld also use codes.get('a') which has the advantage that if tried to access a character that isn't in codes no error will occur.

Let's now write a function called send_morse_for that takes a single letter and will eventually flash the Morse code for that letter. You can find this, in *05_03_morse_send_morse.py*. This function doesn't do any flashing yet, it just uses print commands so that we can see what it does. Eventually, we will flesh out this skeleton of a function so that it flashes the LED; for now, we just want to print out the character chosen and its Morse representation.

```
def send_morse_for(character):
    if character == ' ':
        print('space')
    else:
        dots_n_dashes = codes.get(character.lower())
        if dots_n_dashes:
            print(character + ' ' + dots_n_dashes)
        else:
            print('unknown character: ' + character)
```

Run the program and then enter the following in the Shell:

```
>>> send_morse_for('z')
z --..
>>>
```

Let's take a look at what's going on. First we check to see if the character passed into the function for sending is a space character. If it is, we just print the string `'space'` so we can see the program is working. This will later be replaced by a space-length pause.

If the character is not a space character, then the `codes` dictionary is used to lookup the `dots_n_dashes` to be flashed. We want the program to work for both upper and lower case letters and so, when looking up the dots and dashes in the dictionary, the letter is converted to lowercase using the `lower` method.

If the character is in the dictionary (determined by `if dots_n_dashes`), the `print` command reminds us of the character and prints the sequence of dots and dashes. If, on the other hand, the character is not in the codes dictionary (it might be a number or punctuation) the missing character is printed.

Entering each character separately would definitely start to get a bit tedious after a while, and so let's expand the program so that it lets us type in a whole sentence. You will find this in *05_04_morse_printing.py*. We have added an eternal loop containing code to prompt us to enter some text.

```python
while True:
    text = input('Message: ')
    for character in text:
        send_morse_for(character)
```

Try running it and enter some text, as shown below. From what's being printed out, things look very promising.

```
Message: Morse Code
M --
o ---
r .-.
s ...
e .
space
C -.-.
o ---
d -..
e .
Message:
```

Before we go any further, let's find out a bit more about the timings involved in ITU, the current international Morse code.

A quick search on the web tells us that Morse code does not dictate the length of the dots in seconds, but it does state that the duration of a dash should be three times the duration of a dot. The gap between each letter of a message should also be three times the duration of a dot, and the gap between words should be seven dots-worth of duration. We can use this information to define some variables at the top of our program like this:

```
dot_duration = 0.2 # arbitrary choice
dash_duration = dot_duration * 3
word_gap = dot_duration * 7
```

By defining the dash_duration and word_gap in terms of the dot_duration, it is easy to change the speed of the transmission – should we want to speed up or slow down our sending of Morse code, all we need to do is change dot_duration, and the other time delay times will take care of themselves.

It's time now to move on to some flashing. We can make this happen by first adding the imports at the top of the file for Pin and sleep and defining the LED like this:

```
from machine import Pin, Signal
from time import sleep

led = Signal(22, Pin.OUT, invert=True) # for ESP32 Lite
```

Now, we need to modify the function send_morse_for so that, instead of just printing what its planning to flash, it actually does the flashing. Here is the modified function:

```
def send_morse_for(character):
    if character == ' ':
    sleep(word_gap)
    else:
        dots_n_dashes = codes.get(character.lower())
        if dots_n_dashes:
            print(character + ' ' + dots_n_dashes)
            for pulse in dots_n_dashes:
                send_pulse(pulse)
                sleep(dash_duration)
        else:
            print('unknown character: ' + character)
```

The first thing that happens is that, if the character being flashed is a space (signifying the end of a word), there is no flashing to do —- but we do need to delay for word_gap seconds. On the other hand, if the character is not a space, then we look-up the dot and dash pattern for the character in the codes dictionary.

If the character is not there (maybe the message included numbers or other characters not in the codes dictionary) then we print a message to that effect; if it is there, then we need to flash the pattern of dots and dashes. So, a for loop is used to step over each dot or dash in turn and then call a function (that we haven't written yet) called send_pulse. This new function receives a single '.' or '-' as its parameter, and does a short or long flash depending on whether it's a dot or a dash.

```
def send_pulse(dot_or_dash):
    if dot_or_dash == '.':
        delay = dot_duration
    else:
        delay = dash_duration
    led.on()
    sleep(delay)
    led.off()
    sleep(delay)
```

The send_pulse function sets the value of the variable delay, depending on whether the character is a dot or a dash, and then does a single flash using that delay time.

You can find the full Morse program in *05_05_morse_complete.py*. Run it, and type some messages into the Shell (Figure 5.3). You should now see the message being flashed out on the ESP32 board's built-in LED.

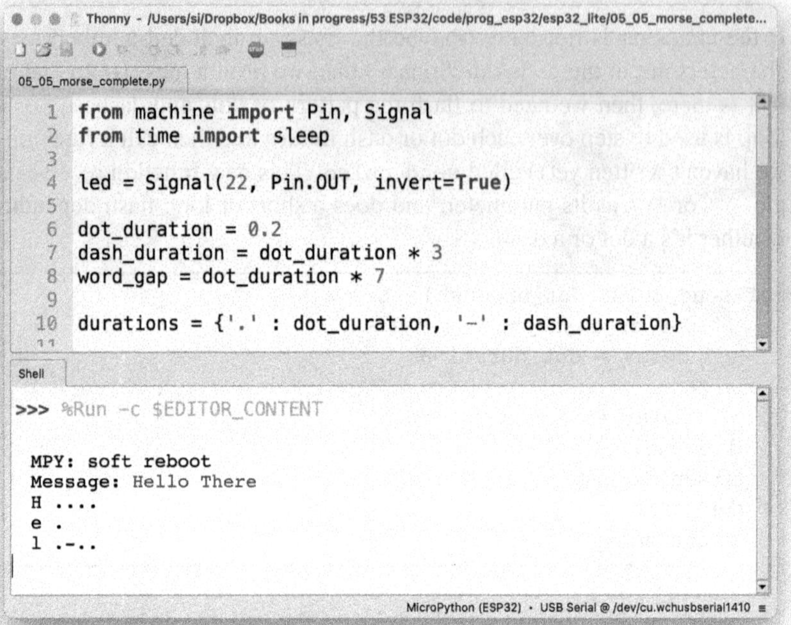

Figure 5.3. The Morse Code Translator in Thonny.

```
from machine import Pin, Signal
from time import sleep

led = Signal(22, Pin.OUT, invert=True)

dot_duration = 0.2
dash_duration = dot_duration * 3
word_gap = dot_duration * 7

durations = {'.' : dot_duration, '-' : dash_duration}

codes = {
    'a' : '.-',    'b' : '-...', 'c' : '-.-.',
    'd' : '-..',   'e' : '.',    'f' : '..-.',
    'g' : '--.',   'h' : '....', 'i' : '..',
    'j' : '.---',  'k' : '-.-',  'l' : '.-..',
    'm' : '--',    'n' : '-.',   'o' : '---',
    'p' : '.--.',  'q' : '--.-', 'r' : '.-.',
    's' : '...',   't' : '-',    'u' : '..-',
```

```
    'v' : '...-', 'w' : '.--', 'x' : '-..-',
    'y' : '-.--', 'z' : '--..'
}

def send_pulse(dot_or_dash):
    if dot_or_dash == '.':
        delay = dot_duration
    else:
        delay = dash_duration
    led.on()
    sleep(delay)
    led.off()
    sleep(delay)

def send_morse_for(character):
    if character == ' ':
        sleep(word_gap)
    else:
        dots_n_dashes = codes.get(character.lower())
        if dots_n_dashes:
            print(character + ' ' + dots_n_dashes)
            for pulse in dots_n_dashes:
                send_pulse(pulse)
            sleep(dash_duration)
        else:
            print('unknown character: ' + character)


while True:
    text = input('Message: ')
    for character in text:
        send_morse_for(character)
```

As an exercise, you might like to try adding in the extra number characters specified in the Morse code (see https://en.wikipedia.org/wiki/Morse_code).

You might also want to try speeding things up by changing the value of dot_duration.

Summary

In this chapter we have learnt how useful lists and dictionaries can be. We could have made our Morse program using a huge series of if statements, checking for each letter in turn and then flashing the dots and dashes, but this would have resulted in a program hundreds of lines long. By using a dictionary, we have kept our program to a nice concise 50 or so lines of code.

In the next chapter, we will look at some more ideas to keep our code neat and tidy.

CHAPTER **6**

Modules, Classes and Files

We have been importing modules into our programs since the start of this book – so, it is finally time to take a look at exactly what's going on here, and introduce the concept of classes and object-oriented programming.

This chapter is getting a little advanced, so don't worry if not all of it makes sense on first reading – you may find yourself coming back to it as you progress through the book.

Importing from Modules

MicroPython includes lots of ready-made functions for us to use in our programs. If every ready-made function were immediately available to us, then there would be a real risk that (when we are thinking up names for our own functions) a function with that name might already exist. If there are two functions with the same name, how will Python know which one to use when you call the function?

Try running the following program:

```
def print():
    pass
print()
```

In Python, `pass` does not actually do anything, but is required in situations, like this, where there would otherwise be an empty block of code.

When you run this, you will see a somewhat cryptic error message like this:

```
ManagementError
```

```
>>> THONNY FAILED TO EXECUTE COMMAND get_globals
SCRIPT:
... lines removed for brevity...
TypeError: function takes 0 positional arguments but 3 were
    given
```

This is Python's way of telling us that we can't give our own function the name `print`, because that name is already in use.

The function `print` is available to use in programs without having to do anything special to include it. Other functions (the vast majority) are hidden away in modules, and we can only use them in our programs by importing them like this:

```
from time import sleep
```

This tells Python that we want to use the `sleep` function from the `time` library. Once we have made the `import` at the top of the program, we can simply call `sleep` anywhere in your programs like this:

```
sleep(0.5)
```

If we were feeling adventurous, we could just import everything from the `time` library in one go by writing:

```
import time
```

The down side to importing everything is that, unless we have an encyclopaedic knowledge of Python, we don't know exactly what functions will be imported.

Documentation

We know `sleep` lives in the module `time`, but we have no idea what other functions we might be importing at the same time with `import time` and, if we only need `sleep`, then it's safest just to import sleep. We can actually see what else is available in `time` either by consulting the documentation for `time` (`https://tinyurl.com/3w2cvej5`) or by running the `dir` command like this:

```
>>> import time
>>> dir(time)
['__class__', '__name__', 'gmtime', 'localtime', 'mktime',
```

```
'sleep', 'sleep_ms', 'sleep_us', 'ticks_add', 'ticks_cpu',
'ticks_diff', 'ticks_ms', 'ticks_us', 'time', 'time_ns']
>>>
```

It's possible to guess what some of these functions do from their names, but to make proper use of them, we may need to consult the documentation. The documentation for MicroPython is available at https://docs.micropython. org/ Be aware, though, that this is not specific to the ESP32 version, and not all modules will be available; equally importantly, the ESP32 specific machine and esp modules are documented separately at https://tinyurl.com/ 2rhd8ahy.

Useful Built-in Modules

MicroPython has a large number of useful modules that can save you time with your programs. There is no sense in reinventing the wheel, especially if someone who is really good at making wheels will give you one for free!

You can find a list of all the modules at the MicroPython documentation here: https://tinyurl.com/29m46xyz.

Note that often the words *library* and *module* are used interchangeably. Also, because this MicroPython is running on an ESP32, memory space is at a premium. If you are used to using full-strength Python, you will find that not everything in every module is implemented. However, all the most useful stuff should be there.

Here are some highlights of the most used built-in modules.

random

In our dice example, earlier in the book, we met the randint function from the random module. We used this function to generate a random number in a range, like this:

```
>>> randint(1,6)
4
>>>
```

Another useful function in the random module is choice. You can use this function to randomly select one of the entries in a list like this:

```
>>> from random import choice
```

```
>>> fruit = ['apple', 'banana', 'orange', 'grape']
>>> choice(fruit)
'apple'
>>> choice(fruit)
'grape'
>>>
```

While you could just pick a random number between 0 and the length of the list minus 1, and then access the element of the list at that index position, using choice is nicer.

math

You get all the basic arithmetic operators like +, -, * and / (and some others) without having to use a module. However, if you need more advanced functions from trigonometry, logarithms etc., you have to use the math module.

The mathematical constants e and Π are available.

```
>>> from math import pi
>>> pi
3.141593
```

However, it is important to know that all the trigonometric functions in the math module use radians rather than degrees. There is a handy function radians that will do the conversion from degrees to radians. So, for instance, to find the sin of 30 degrees you could do this:

```
>>> from math import radians, sin
>>> sin(radians(30))
0.5
>>>
```

Another math function that is quite useful is pow. This raises one number (int or float) to the power of another. For example:

```
>>> from math import pow
>>> pow(2, 16)
65536.0
>>>
```

There are many more functions in this module and, if you find that there is something mathematical that you need to do, check first—there will probably already be a function for it.

Classes

Python is (like most recent languages) object-oriented. There are varying degrees of *object-orientedness* in languages and, in the case of Python, it's not radically object-oriented. Let's look at what this means.

If you recall, back in Chapter 5, we wanted to prepare messages to be sent by Morse code by converting them to lowercase. We did this using the method `lower()` like this:

```
>>> 'AbC'.lower()
'abc'
>>>
```

A lowercase string with the same letters has been created for us from the mixed case string `'AbC'`. However, you may be asking yourself why the dot, and why that order. Why aren't we treating `lower` as a function and writing the following (which won't work by the way)

```
>>> lower('AbC')
```

especially as we can write:

```
>>> len('AbC')
3
>>>
```

This really highlights Python's fairly lackadaisical approach to object-orientation – sometime you use the dot notation, sometimes you use a function call. So, lets take a look at what's going on with the object-oriented way of doing things.

```
>>> 'AbC'.lower()
'abc'
>>>
```

The formal way we would describe the use of the dot here is to say that we are calling the method `lower` on the *instance* of the *class* string `'AbC'`. Let's unpack that and define a few things.

First of all *strings*. There might be lots of strings in use in a program, and they all have one thing in common – they consist of a list of characters. You could say that they are all of the same type or *class*. In fact, we can ask Python to tell us the class of something like this (note that's double underscore before and after the word class)

```
>>> 'abc'.__class__
<class 'str'>
>>> [1,2,3].__class__
<class 'list'>
>>>
```

The string 'abc' is said to be an instance of the class str – and the thing about a class is that, as well as saying something about the type of data it contains, it can also contain functions that are particular to that class (such as lower in this example).

Functions like lower that are linked to a class are called *methods*. We can see all the methods available on the class str like this:

```
>>> dir(str)
['__class__', '__name__', 'count', 'endswith', 'find', 'format',
'index', 'isalpha', 'isdigit', 'islower', 'isspace', 'isupper',
'join', 'lower', 'lstrip', 'replace', 'rfind', 'rindex', 'rsplit',
'rstrip', 'split', 'startswith', 'strip', 'upper', '__bases__',
'__dict__', 'encode']
>>>
```

As you can see, there's lower and the corresponding upper and a load of other methods – some of which are more useful than others.

Modules often contain a class with methods defined in it. We have already met this in the form of the machine library, from which we imported Pin and Signal (both of which are classes) in order to be able to make an LED blink. Here's a reminder, where we have simplified program *03_01_blink* just to turn the LED on. If you don't want to type this in, you can find it in the file *06_02_Pin_class.py*.

```
from machine import Pin, Signal
led = Signal(22, Pin.OUT, invert=True)
led.on()
```

You may wonder why Signal has a capital S and Pin has a capital P. That's be-cause Signal and Pin are the names of classes and class names, by convention, start with an uppercase letter. So, when we import Pin, we are not importing a single function like we did when we imported randint from random, but, we are importing an entire class called `Pin`.

We can find out what `Signal` is capable of (what's in it) by doing this:

```
dir(Signal)
['__class__', '__name__', 'value', '__bases__', '__dict__', 'off',
    'on']
```

The methods `on` and `off` are there, and some other things that we don't need to know about, as well as some others that we will meet in later chapters.

You can define your own classes and methods, but you rarely need to do this when writing small programs for the ESP32, so we will consider this an advanced topic and leave it for other resources to cover. If you are interested, you can find a good explanation of classes in Python at `https://tinyurl.com/msdkvys5`.

String Methods

In uncovering how our use of the `lower` method worked, we have discovered other methods that we can use on strings, some of which are very useful. While we are here, let's go through some of these.

find

This method returns the index position for the starting point of one string within another. If the string being sought isn't in the string being searched, then this method returns −1. For example

```
>>> 'ESP32 Lite'.find('Lite')
6
>>>
```

In this example we are searching for the string `'Lite'` within the string `'ESP Lite'` and need to remember that the index starts from 0 not 1. This method is often used to find out if one string contains another, or not, by seeing if it returns −1.

Formated String Literals

Whether displaying messages in the Shell, or perhaps on a display attached to a ESP32, a common problem is the need to display a combination of a message and some values. MicroPython has a handy way of doing this called *f-strings*. f-strings are prefixed by a f before the opening quote, and contain Python commands bracketed between curly braces. For example:

```
>>> temperature = 23.4
>>> f'The temperature is {temperature} degrees C'
'The temperature is 23.4 degrees C'
>>>
```

You can substitute any Python expression you like between the {} markers – you are not restricted to just variable names.

You can also achieve the same result by concatenating strings using +. However, each thing that you add must be a string, and often numbers will need to be converted into strings using the str function.

```
>>> 'The temperature is ' + str(23.4) + ' degrees C'
'The temperature is 23.4 degrees C'
```

Which approach you take is entirely a matter of personal taste.

replace

You can create a new string from an existing string, with some of the text replaced using the replace method. For example, if you wanted to change all occurrences of *b* in a string to *B* you could do it like this:

```
>>> 'abcb'.replace('b', 'B')
'aBcB'
```

split

If you have a string that contains a file path such as: */home/pi/myfiles/file1.txt*, and you want each part of the file path as its own entry in a list, then the split command will save you a lot of effort.

In the example, the file path is first split on the / character to find its parts. The last part (the actual file name) is then split on . to separate the base file name from the extension.

```
>>> parts = '/home/pi/myfiles/file1.txt'.split('/')
>>> parts[:-1]
['', 'home', 'pi', 'myfiles']
```

```
>>> file = parts[-1]
>>> file
'file1.txt'
>>> file.split('.')
['file1', 'txt']
```

Notice how [-1] is used as a shorthand to provide all the list elements up to (but not including) the last one, and [-1] picks out the last one only.

strip

A common problem when dealing with a string that a user has typed in -- perhaps from an `input` command – is that there are unwanted spaces on either end of what they typed. The `strip` method removes those spaces without touching any spaces between words in the string. Here's an example that assumes the user has typed in their name is a rather sloppy manner.

```
>>> ' Simon Monk '.strip()
'Simon Monk'
```

Files

The ESP32, has some 2MB of flash storage, that has to be shared between the program running on the ESP32 and any data that you might want to store on the ESP32. 2MB might not sound like very much but, for a small microcontroller board, this is well above average – which means that you have some space available for storing data on the ESP32 itself. This can be very useful in projects like data-loggers, where you store sensor readings to a file periodically and then, when the logging is finished, copy the file of readings onto your computer.

Being able to save data persistently to a file, and then read it back from the file later, is also very useful when you want your program to remember things after a reset. For example, if you had a program that operated in different modes (perhaps turning something on and off) you might want to remember what state it was in when the ESP32 was powered down. Another example would be to remember a high-score in a game.

We will start by creating a file called *test.txt*, and writing the number 1 into it. You can find this program in *06_03_file_write.py*.

```
f = open('test.txt', 'w')
```

```
f.write(str(1))
f.close()
```

Run the program and then, in the Thonny *View* menu, enable *Files*. You should now see two lists on the left of the Thonny window – one for files on your computer and one for files on the ESP32. You should see the file *test.txt* there (see Figure 6.1).

Figure 6.1. The files view in Thonny.

If you click on the file *test.txt* to open it, you should see that it does indeed contain the number 1. It is important to remember that the file is actually on your tiny little ESP32, not on your computer. You can edit the files, so change 1 to 2 and then, click the *Save* icon – your file will now contain the number 2.

Now that we have a file on the ESP32 to use, let's write a program to read it. You can find this in *06_04_file_read.py*.

The file is opened in write mode (using the parameter `'w'` written to using the method `.write()`, and closed again using the method `.close()`. We can only write strings into this text file, and so the number 1 has to be converted into a string using `str`.

```
f = open('test.txt', 'r')
print(f.read())
f.close()
```

This time, the files is opened in read mode using the parameter ('r'). You should see output in the Shell like this:

```
>>> %Run -c $EDITOR_CONTENT

MPY: soft reboot
2
>>>
```

Exceptions

Reading a file like this is fine if we know that the file already exists – but a common scenario in reading a file on the ESP32 is that, the very first time the program is run, an attempt is made to read a non-existent file. If the file doesn't exist, then we probably want to create it rather than have an error occur.

The way to do this in MicroPython is to try to read the file and, if we get an error, then we know the file doesn't exist. So far, when we have had errors (such as when accessing a dictionary with an unknown key) the program has stopped (crashed) because of the error. In MicroPython we can catch such errors when they occur, and handle them in our code without it causing the program to crash. To do this, we use try ... except, as the example in *06_05_try_except.py*.

```
file_name = 'testXX.txt'
try:
    f = open(file_name, 'r')
    print(f.read())
    f.close()
except:
    print('File {} doesn't exist'.format(file_name))
```

Notice that the filename has been changed to *testXX.txt* and, because there is no such file on the ESP32, we will see the message to that effect. The point is that the program did not crash, we are catching any error using the try ... except statement and handling the error ourselves. Try changing *file_name* to *test.txt*, rerun the program, and notice the difference.

File Counter Example

Putting the read and write file programs together, the program *06_06_counter.py* will keep track of how many times the ESP32 has been reset and its program run.

```
config_file = 'config.txt'
count = 0

def read_config():
    global count
    try:
        f = open(config_file, 'r')
        count = int(f.read())
        f.close()
    except:
        pass

def write_config():
    f = open(config_file, 'w')
    f.write(str(count))
    f.close()

read_config()
print(count)
count += 1
write_config()
```

The program starts by defining two variables. The name of the file is held in `config_file` – this idea of a configuration file is common in embedded programming.

The other variable (`count`) will be used to hold the count of times the program has been run.

Inside the `read_config` function the line `global count` indicates to Python that the variable `count` is defined globally, outside of the function. You can read global variables from within a function (as we do in the `write_config` function) but, if you want to change their value, you need a `global` line like this at the top of the function.

The statement below `except` is just `pass`. That is needed because Python insists on there being something inside `except` – if you don't want to do anything, you indicate this by using `pass`.

After the two function definitions, we have the code that will be run just once before the program finishes. This reads the config file, prints the value of `count` that it reads, adds one to count, and then saves it again.

Run the program a few times and you will see the count increse by one each time.

If you now look in the file section of Thonny (after clicking *Refresh* in the little drop-down menu in the file area) you should see a new file *config.txt* and, inside it, you should see the current count.

Summary

In this chapter we have learnt a bit more about modules, and how to use them, as well as the sometimes confusing style that results from Python's implementation of object-oriented programming. We have also explored the use of the ESP32 board's spare flash memory for use as a file system.

In the next chapter, we will start using your ESP32 board's GPIO pins.

CHAPTER 7

Inputs and Outputs

In this chapter we will look at using the ESP32 board's GPIO pins. Rather than wade into the depths of electronics, we will restrict ourselves to nothing more complex than a metal paper-clip (or short length of wire) and the ESP32 board's built-in LED. This way, we can be ready to embrace some proper electronics when we come to Chapter 8.

Digital Outputs

We have already used digital outputs quite extensively in controlling the built-in LED. This LED is connected to pin 22 of an ESP32 Lite board, or pin 2 of an ESP32 DevKit 1 board. For most boards, this pin is also brought out to one of the connections running down the sides of the board. The voltage of the pin can be set to be either 0V or 3.3V. If you have something external (say an LED) connected to that pin, then you can turn it on by setting the pin to 3.3V or off by setting it to 0V. It is actually a bit more complicated than this, as we will discover in Chapter 8 when we come to connect an external LED to a pin, but this will do for now.

The reason the pins are called *GPIO pins* is that they are *general purpose* – that is, they can act as inputs or outputs. Before we can use a pin as a digital output, we need to tell the ESP32 that this is how we intend to use it, like this:

```
from machine import Pin, Signal
led = Signal(22, Pin.OUT, invert=True)
```

As we discussed in Chapter 6, Pin is a class and, OUT has a value that tells MicroPython that the pin is to be used as an output. Now we have a reference to the pin 22, and we have told the ESP32 to configure it as an output.

The `Signal` class is useful in that it gives us the methods on and off that will behave correctly irrespective of how the built-in LED is wired up. (As long as we have set `invert=True` if the LED logic is inverted for our board.)

As well as the methods on and off we can also use the method value that allows us to set the output voltage.

```
>>> led.value(1)
>>> led.value(0)
```

In fact, if we don't mind just using value instead of on and off, and are in control of our own logic as far as on and off goes, we can dispense with `Signal` altogether and just define an output pin directly like this:

```
from machine import Pin
led = Pin(22, Pin.OUT)
```

> TIP: In the Shell, pressing your up arrow cursor key will recall the last command you entered in the Shell, so that you don't have to type the whole line again.

Digital inputs

Whereas for a digital output, the ESP32 board is controlling something, turning it on or off, a digital input is used to detect whether a switch of some sort is on or off. A digital input is *read*, to see whether the voltage at the pin is above or below a threshold roughly half-way between 0V and 3.3V.

You should not be surprised to hear that this is how we tell a ESP32 board to use a particular GPIO pin (12 in this case) as an input:

```
>>> switch = Pin(12, Pin.IN)
```

A convention, when connecting switches to a digital input, is to use what is called a pull-up resistor (Figure 7-1) connected to the digital input, that biases the input towards 3.3V (high). The switch is then connected between the digital input and *GND* (0V), so that when the switch is pressed the digital input is connected to GND. Think of the pull-up resistor as a spring that pulls the GPIO pin high unless pulled the other way more strongly by the switch.

This means that when the switch is not pressed the digital input is at 3.3V (high) and when the switch is pressed, the input becomes 0V (low). This can be confusing as, logically, high usually means *on* rather than *off* – but, that's fine, we can just compensate for this in the code.

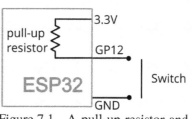

Figure 7.1. A pull-up resistor and digital input.

We don't actually have to use a physical pull-up resistor when connecting a switch, because the ESP32 board's GPIO pins all have one built-in that can be enabled and disabled from your code. So, when connecting a switch to a ESP32 board you write the following – both to set the pin as an input and enable the built-in pull-up resistor.

```
switch = Pin(12, Pin.IN, Pin.PULL_UP)
```

The optional third parameter Pin.PULL_UP switches in the pull-up resistor.

We can test out digital inputs by running the program below (*07_01_digital_input.py*).

```
from machine import Pin
from time import sleep

switch = Pin(12, Pin.IN, Pin.PULL_UP)

while True:
    print(switch.value())
    sleep(0.1)
```

The listing above is for the ESP32 Lite, and pin 12 is chosen because it's next door to a GND pin, so we can simulate a switch just by connecting the two together with a paperclip or other conductive metal object. The ESP32 DevKit 1 version of the program uses pin 15 rather than 12, because on the DevKit 1, pin 15 happens to be next to a GND pin.

When you see a stream of 1s appear in the Shell. This is because the input is being pulled-up *high* — here 1 means high (3.3V or at least half-way to 3.3V and definitely not 0V).

`switch.value()` is used to determine whether the input is high (1) or low (0). Now, as shown in Figure 7.2, use an unfolded paper-clip, screwdriver or other metallic object to connect GPIO 12 to the convenient GND pin just above it. You should see that you now get 0s in the Shell.

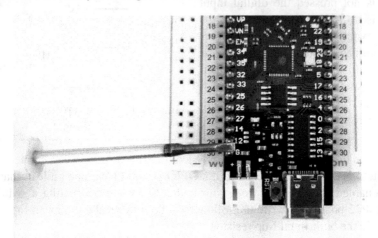

Figure 7.2. A make-shift switch using a screwdriver.

You can use `Signal` with a digital input (rather like with a digital output) to invert the logic, so that connecting a pin to GND gives a `value` or 1 rather than 0.

```
from machine import Pin, Signal
switch = Signal(22, Pin.IN, Pin.PULL_UP, invert=True)
print(switch.value())
```

However, I think, other than the useful case of the built-in LED (the logic of which is not under our control) it's fine to just use `Pin` on its own when it comes to digital inputs.

Let's now look at another example of using a digital input. The program *07_02_switch_led.py*, will light the built-in LED for 10 seconds when our makeshift switch connects GPIO 12 to GND.

```
from machine import Pin, Signal
from time import sleep
```

```
switch = Pin(12, Pin.IN, Pin.PULL_UP)
led = Signal(22, Pin.OUT, invert=True)

while True:
    if switch.value() == 0:
        led.on()
        sleep(10)
        led.off()
```

Try running the program. Initially, the built-in LED will be off, but when you momentarily connect GPIO 12 and GND using the paperclip, the LED will light and stay lit for 10 seconds.

A paperclip or screwdriver is not an ideal way of making a switch and, in the next chapter, we will explore the use of proper push-switches connected to digital inputs.

Analog Outputs

Digital outputs allow us to turn things on and off – for example, we can light an LED or turn it off – but we can't use a digital output to change the brightness of an LED in a graduated way. That is where analog outputs come into play.

Analog outputs on the ESP32 board should, more accurately, be described as Pulse Width Modulation (PWM) outputs. They are actually digital outputs that control the length of pulses that they generate. This can be used to control the brightness of an LED or the speed of a motor. Figure 7.3 shows how PWM works, where the various percentages are the amount of time that the signal is high.

Think, for example, of an LED. If the repeating pulses are quite short, then the LED is only lit for a short amount of time and will appear to be dim. If the pulses are longer, then the LED will appear brighter.

Since the pulses are so fast (by default 5000 pulses per second) your eye can't keep up, and you can't see the LED flickering.

You can use PWM on any of the ESP32 board's pins, so let's try controlling the brightness of the built-in LED. Try running the program *07_03_pwm.py*.

```
from machine import Pin, PWM

led = PWM(Pin(22))
```

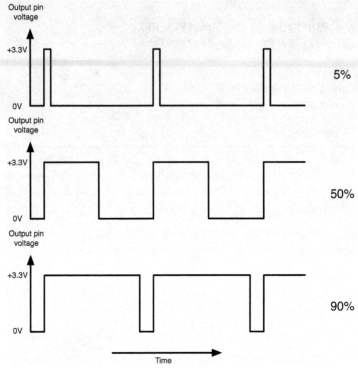

Figure 7.3. Pulse Width Modulation.

```
while True:
    brightness_str = input('brightness (0-65535):')
    brightness = int(brightness_str)
    led.duty_u16(brightness)
```

This program prompts you to enter a number in the Shell between 0 and 65535. Try entering some numbers and notice how the brightness of the built-in LED varies.

```
brightness (0-65534):0
brightness (0-65534):65535
brightness (0-65534):32000
brightness (0-65534):50000
brightness (0-65534):0
brightness (0-65534):
```

Notice that the higher the number, the dimmer the LED (if you are using an ESP32 Lite). That is because the LED is actually off when the pin is at 3.3V.

(Remember the on/off logic of the LED is inverted). The opposite happens when using the ESP32 DevKit.

When specifying how long the pulse is high (the duty) we enter a number between 0 and 65536. This number is 2 raised to the power of 16, minus 1. This is because a 16 bit number is used to control the PWM pulse duration.

When specifying that the built-in LED is to be used for PWM output, we wrap the Pin definition in a PWM class – remember that the input function returns a string, so this has to be converted to the number expected by duty_u16 using the int function. The function duty_u16 is so called because it expects a 16 bit unsigned number.

Over 65,000 steps of brightness is excessively precise for an application like this, so we could modify the program to that of *07_04_pwm_percent.py*, where we enter a number between 0 and 100% brightness.

```
from machine import Pin, PWM

led = PWM(Pin(22))

while True:
    brightness_str = input('brightness (0-100):')
    brightness = int(int(brightness_str) * 65535 / 100)
    led.duty_u16(brightness)
```

Note that we have had to use int twice. Once to convert the text entered into a number and then again to convert the floating point number resulting from the multiplication and division into the integer value expected by duty_u16.

Generating pulses like this can also be used to control servomotors. We will see how to do this in the next chapter.

Analog Inputs

Digital inputs from something like a switch are either on or off. However, what if we want to measure the voltage at the pin, rather than simply detect if it is over the threshold voltage or not. Such a voltage might come from a sensor that we are using to measure temperature or light.

When you take a digital reading, you only get one of two possible values (0 or 1). However, with an analog reading on an ESP32, you get a reading that is a number between 0 and 65535. The default voltage range for analog inputs on an ESP32 is between 0 and 1.0V, not the 3.3V you might be expecting. That means that a voltage of 0V will give a reading of 0 and a voltage of 1.0V should give a reading of 65535.

Figure 7.4 shows a plot of the analog reading that you will get (y-axis) against the voltage at the analog input pin (x-axis). These readings are the average of 100 readings at each voltage, and the error bars show +/- three standard deviations. That is, any single analog reading should fall within this range. As you can see, there is quite a lot of variation on the readings so, despite the precision of the reading, the graph is not hugely consistent. It is however, fine for getting a rough idea of say the light level, or some other measurement from an analog sensor.

You can also see that there is a bit of a *dead* area near 0V, and the readings only actually start to increase after about 0.05V.

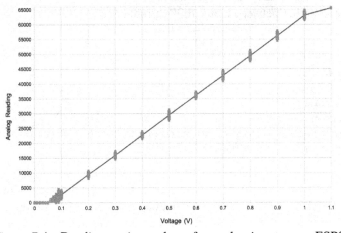

Figure 7.4. Reading against voltage for analog input on an ESP32.

Try running program *07_05_analog_input.py* and you should see a stream of numbers appear in the Shell. There will be some fluctuation as the analog input will be picking up electrical noise from the pin which is not connected to anything (called *floating*).

```
from machine import ADC, Pin
from time import sleep

analog = ADC(32)

while True:
    reading = analog.read_u16()
    print(reading)
    sleep(0.5)
```

Just like we wrapped `Pin` in `PWM` to use PWM, when we use an analog input the pin must be wrapped in the `ADC` (Analog Digital Converter) class. Note that, unlike digital inputs and outputs, analog inputs are NOT available on every pin. You can use any pin, except pins: 16 (RX2), 17 (TX2), 5, 15, 23, 19 and 22 – but you wouldn't be able to use 22 on an ESP32 Lite anyway, because the built-in LED is attached to that pin. For the same reason, you can't use pin 2 as an analog input on a ESP32 Devkit 1.

If you are using the ESP32's WiFi or Bluetooth features then some of the analog input pins (0, 2, 4, 12-15 and 25-27) also become unavailable, so it's a good idea to stick to pins 32-39 when it comes to analog inputs.

Summary

This chapter has dealt with the basics of inputs and outputs, but the ESP32's input/output features are actually very advanced so, in Chapter 12, we will return to the subject of input and output and look at some of its more advanced features.

In Chapter 9, we will return to analog inputs and show how you can attach sensors to them.

In the next chapter, we will start using the ESP32 with some external electronic components, and learn how to make simple electronic add-ons using solderless breadboard.

CHAPTER 8

Electronics

This chapter builds on what you have learnt about analog and digital inputs and outputs, and uses this knowledge to construct some simple projects using a solderless breadboard to connect electronic components to your ESP32 board (Figure 8.1).

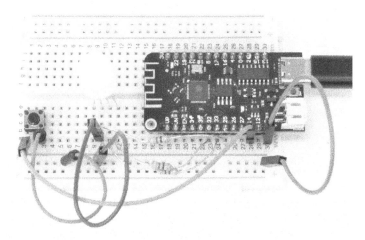

Figure 8.1.   An example project using solderless breadboard.

Solderless Breadboard

Solderless breadboard is a great way to try out electronic circuits. The components (including the ESP32 board) are pushed into metal clips behind the plastic face of the breadboard, so there is no need for any soldering – and it is easy to make different projects reusing the same components in different configurations.

Solderless breadboard is available in various sizes. In this book we use size called *400 point*, or *half-sized*. These are ideal for holding an ESP32 board and a few extra components.

As you can see in Figure 8.1, an ESP32 board is plugged into the solderless breadboard, along with some other components that we will discuss later. Jumper wires are used to connect different parts of the circuit together.

Behind the face of the solderless breadboard (let's just call it breadboard from now on) you will find metal clips (Figure 8.2) that connect together (electrically) any component legs that are pushed through from the front.

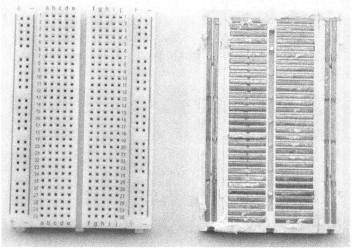

Figure 8.2. Breadboard deconstructed (front view left, back view right).

The main area of the breadboard is divided into two columns of clips that connect all the holes of a particular row together. At the left and right side of the breadboard are long vertical clips called power buses. These can be used for any connections that you want to make, but are particularly useful for GND and 3.3V power connections that often need to be connected to lots of places on the breadboard – and make it much less likely that you will make a connection to the wrong pin. These are marked with a red + and a blue – sign, for clarity.

Each row of the breadboard is marked with a number (0 to 30) and each column with a letter (a to j), helping to identify positions on the breadboard.

The extra width of the ESP32 DevKit 1 leaves only the pins of one side of the board accessible on a standard breadboard. However, solderless breadboard will normally clip together, so with two breadboards, you can often arrange them so that an ESP32 DevKit 1 will span the two boards, making all the pins accessible.

Components

Although programming in Python can be a lot of fun, there is even more fun to be had when you start connecting extra electronics components to your ESP32 board. Components with legs (called through-hole components) were originally designed to be soldered through holes in the circuit board; however, most circuit board designs (like your ESP32 board) now use *surface mount* components. Fortunately, through hole components are still readily available and are great for home projects and making prototypes using breadboard. Figure 8.3 shows a selection of useful components.

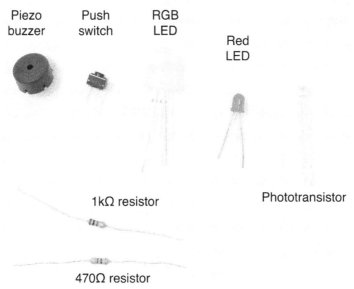

Figure 8.3. Useful Components.

The MonkMakes *Project Box 1 for Raspberry Pi* includes a breadboard and useful set of components to get you started with your electronics projects. You can find out more details here: `https://monkmakes.com/pi_box_1`. You will also find starter component kits and solderless breadboard available from online market places, some of which will provide you with a lot of components for little cost.

Let's look at the components shown in Figure 8.3:

- Piezo buzzer – makes a buzzing sound when connected to an oscillating GPIO pin

- Push switch – the two pins connect electrically when the button is pressed; these are usually connected between a GPIO pin (acting as a digital input) and GND

- RGB LED – a multi-color LED that can be used with three PWM outputs to mix different colors of light by varying the red, green and blue intensity

- Red LED

- Phototransistor – a light-sensitive device that can be used, in combination with a resistor and a GPIO pin (acting as an analog input) to measure light intensity

- 1kΩ resistor – used to convert a current through a phototransistor to a voltage that can be measured using an analog input

- 470Ω resistor – used to limit the current flowing through an LED, whilst still making the LED fairly bright

- 10kΩ variable resistor – often called a *pot or potentiometer*, the position of the knob altering the value of the resistor

We will meet many of these components, and learn how to use them, as we progress through the book.

Making an LED Blink

Having made the ESP32 board's built-in LED blink, lets now attach an external LED to the ESP32 board and make that blink instead. Figure 8.4 show a schematic diagram for this.

We are going to connect the LED to pin 32. When 32 is high (3.3V) current will flow through the resistor (R1 – zig-zag line) through the LED, and back into the ESP32 board's GND connection – making a circuit that causes the LED (D1) to light as the current flows through it. The resistor is needed to reduce the current flowing through the LED – too much current though the LED shortens the LED's life.

Note that, in this book, I have used the American symbol for a resistor. In many parts of the world a resistor is represented just by a rectangle.

In this example, we happened to use a red LED, but any color will work ok.

Connect the components to the breadboard as shown in Figure 8.5. Note that

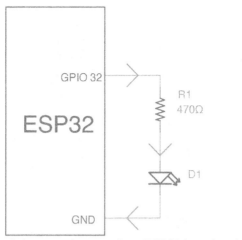

Figure 8.4.   Schematic diagram for a ESP32 board and LED.

the LED has a positive and negative lead – the positive lead is a bit longer than the negative lead and should go to the row also connected to the resistor.

The 470Ω resistor has yellow, violet and brown stripes (ignore the gold one, which only tells you its quality rating) – it does not matter which way around this goes. In actual fact almost any value of resistor between 330Ω and 4.7kΩ will work ok, but the higher the value of resistance (measured in Ω or kΩ = 1000Ω) the dimmer the LED will be.

There are layouts for both the ESP32 Lite and the ESP32 Devkit 1. Note that because the Devkit 1 is a bit wider than the ESP32 Lite, the pins on the right-hand side are not accessible as they have to go in the right-hand column of the breadboard's main area. However, this doesn't really matter, as there are enough pins for us to use on the left. If you have a different ESP32 board from one of these two, then you will need to work out the layout for yourself. If your DevKit 1 is in the wrong place, being careful not to bend the pins, pull it out and reposition it.

Load the program *08_01_blink_led.py* into Thonny and run it. The LED should start to blink. If it doesn't, check your wiring and make sure the LED is the right way around – if the LED happens to be the wrong way around, it won't do it any harm, it just won't light up.

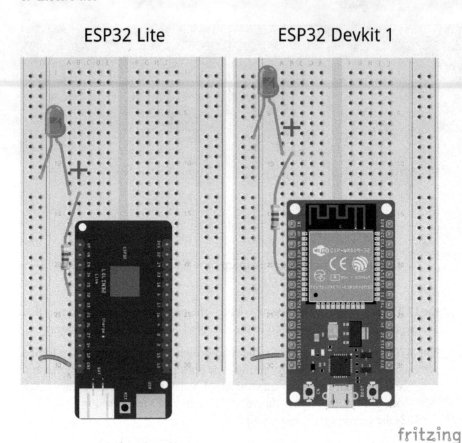

Figure 8.5. The Breadboard Layouts for an LED.

```
from machine import Pin
from time import sleep

led = Pin(32, Pin.OUT)

while True:
    led.value(1)
    sleep(0.5) # pause
    led.value(0)
    sleep(0.5)
```

The code is very similar to what we used to blink the built-in LED, but we are now specifying pin 32 (external LED) instead of 22 (the built-in LED).

RGB LEDs

RGB (red, green, blue) LEDs are made up of three LED colors in one LED body. By varying the amount of red, green and blue (using PWM) it is possible to change the color of the LED. We will start with a simple program that cycles through different colors, and then expand the project to include a push switch that changes the color.

Figure 8.6 show the schematic diagram of our circuit, where R1, R2, R3 are three resistors. Given that LEDs usually have two legs each, you might expect the RGB LED package to have six legs - whereas it has only four. This is because the negative ends of the red, green and blue LEDs are connected to the same pin on the package (the longest leg). Note that some RGB LEDs have all the positive connections in common, so when buying an RGB look for one described as *common cathode*.

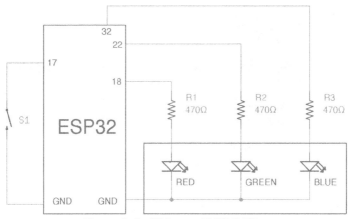

Figure 8.6. Using an RGB LED.

Each LED channel must have its own resistor to limit the current the current; the push switch S1 connects GPIO 17 to GND when the button is pressed.

Before wiring up this (or any complicated) project it's a good idea to unplug the breadboard from USB, so that it isn't powered. This means that nothing will get damaged if you accidentally create a short-circuit.

Start by connecting up the components as shown in Figure 8.7. All three of the resistors used in this project are the same value of 470Ω. You should find that one of the RGB LED's leads is longer than the rest – this is the common negative lead.

For the first part of the project, we won't wire up the switch.

Place the RGB LED such that the long lead is the second lead down (linked to

ESP32 Lite   ESP32 Devkit 1

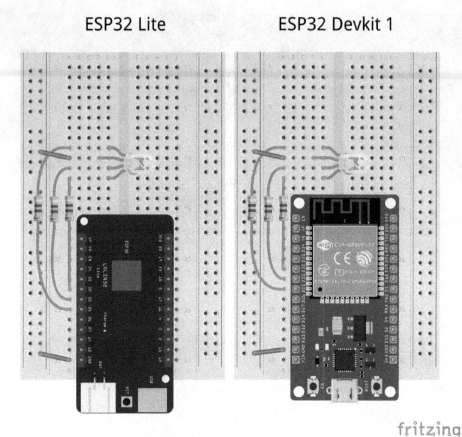

Figure 8.7. The Breadboard Layouts for an RGB LED.

the ground column of the breadboard) and use the three resistors to make connections. The single LED lead next to the longer LED lead is the red channel and should be on row 9 of the breadboard with a resistor going to GPIO 25. The other leads of the LED go to rows 11 and 12. When all the resistors are in place, make sure that the resistor leads are not making contact with anything that they shouldn't, before you reconnect the ESP32 board.

The program for the RBG LED can be found in the file *08_02_RGB.py*.

```
from machine import Pin, PWM
from time import sleep

red_ch = PWM(Pin(25))
green_ch = PWM(Pin(33))
blue_ch = PWM(Pin(32))
```

```
colors = [
    [255, 0, 0],   # red
    [127, 127, 0], # orange
    [0, 255, 0],   # green
    [0, 127, 127], # cyan
    [0, 0, 255],   # blue
    [127, 0, 127]  # purple
]

def set_color(rgb):
    red_ch.duty_u16(rgb[0] * 256) # 16 bit from 8 bit
    green_ch.duty_u16(rgb[1] * 256)
    blue_ch.duty_u16(rgb[2] * 256)

index = 0
set_color(colors[index])
while True:
    index +=1
    if index >= len(colors):
        index = 0
    sleep(0.2)
    set_color(colors[index])
```

This code makes use of a list of lists (colors) to represent the different colors that the LED displays. Each of the inner lists is a list of three values, in order – the red, green and blue brightness, as a number between 0 (off) and 255 (fully on). For example [127, 0, 127] means red half brightness, green off and blue half brightness – making a kind of purple color.

The set_color function takes such an array and sets the three PWM dutycycle values – at the same time, multiplying the value by 256, so that it spans the full range of PWM brightness.

The eternal loop cycles through the colors by adding 1 to index. Once index becomes equal to or greater than the number of colors in the list colors, then we need to wrap around to the start of the list by setting index to 0. Finally include a sleep for 0.2 of a second, to stop the colors changing too quickly. Now run the program, and watch the color of the RGB LED - is it doing what you expect?'

Now let's add a switch connected to GPIO 14 that will change the colors when pressed. The breadboard layout of Figure 8.8 shows the addition of the switch.

To make use of the switch, we need a new program (08_03_RGB_switch.py). Rather than list the whole program, let's just highlight the differences. You may wish to open 08_03_RGB_switch.py in Thonny and follow along.

ESP32 Lite   ESP32 Devkit 1

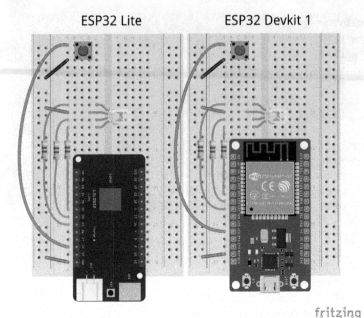

fritzing

Figure 8.8. The Breadboard Layouts for an RGB LED and switch.

First we need to setup a pin as an input for the switch like this:

```
button = Pin(14, Pin.IN, Pin.PULL_UP)
```

The only other change is to the eternal loop, which must now check if the button is pressed (if `button.value()== 0:`)

```
while True:
    if button.value() == 0:
        index +=1
        if index >= len(colors):
            index = 0
        sleep(0.2)
```

Even though the colors are no longer cycling automatically, the `sleep(0.2)` delay prevents switch bouncing, when pressing the switch once may register as several quick presses, as the switch makes repeated contacts as the button is being pressed. This stops some color values appearing to be skipped when the button is pressed. It also means that, if you hold the switch, the LED will cycle through the colors.

Servomotors

Servomotors (Figure 8.9) differ from most motors, in that their intended use is to set the position of an arm attached to their rotor to a particular angle (roughly in the range 0 to 180 degrees) rather than to rotate continuously.

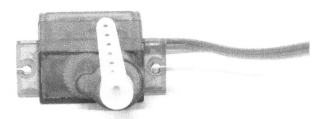

Figure 8.9. A small servomotor.

A servomotor is actually quite a sophisticated little device – Figure 8.10 shows a representation of what's inside one.

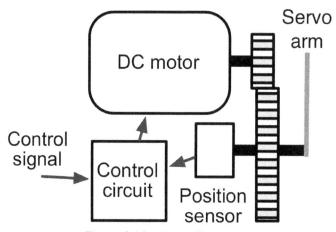

Figure 8.10. A small servomotor.

A DC motor (a normal motor, that rotates continuously) is attached to a gear box, which drives the servo arm. This is also coupled to a position sensor on the output, that provides feedback to an electronic control circuit, so that the servo arm is kept at the correct angle.

Control of the position uses a series of pulses arriving every 20 milliseconds. Figure 8.11 shows how different pulse lengths result in the servo arm moving to a different position.

A short pulse, of just 0.5 milliseconds, will put the arm at one end of its travel.

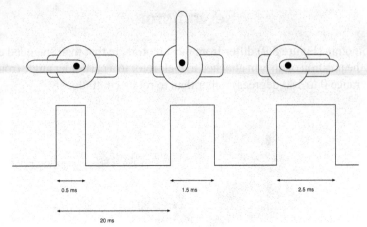

Figure 8.11. Servomotor control pulses.

A pulse of 1.5 milliseconds will put the arm at its center position, and a pulse length of 2.5 milliseconds will put it at the other end of its travel.

Before we look at how we are going to write code to generate these pulses, let's wire up a circuit to experiment with. Figure 8.12 shows the schematic for our test circuit.

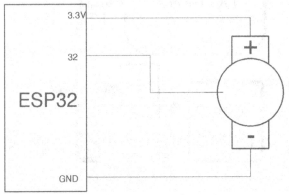

Figure 8.12. The schematic for connecting a servomotor.

A lot of small servomotors (including the one in the MonkMakes kit) will work at 3.3V, but not all servomotors can – so check that it will work at 3.3V before buying. Figure 8.13 shows the breadboard layouts. Note that the position of the ESP32 Devkit 1 has been moved compared to the previous diagrams, so that the pins on the tight-hand side of the board are accessible. When it comes to taking an ESP32 board off a breadboard it is very easy to bend or even break the pins, so carefully lever the board away from the breadboard at each end in turn, a little bit at a time, until the board becomes free.

ESP32 Lite   ESP32 Devkit 1

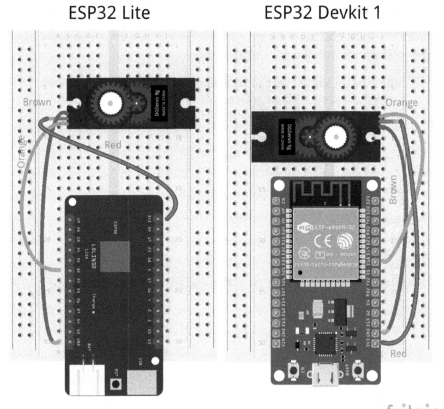

fritzing

Figure 8.13.   The wiring diagrams for a servomotor.

To connect the servomotor, use jumper wires. The red lead of the servomotor should be connected to 3V on the ESP32 board, the brown lead of the servo-motor to GND, and the orange control lead to GPIO 32 or 18 depending on which board you are using – as shown in Figure 8.12.

The servomotor is supplied with a little bag of arms that fit over the cogged drive shaft of the servo motor. Select the one shown in Figure 8.8, and push it onto the drive shaft. It doesn't matter at what angle you attach it, as you can always adjust it later.

Open the file *08_04_servo.py* in Thonny, and run it. You should find that the servo arm will do a little wag left, then right, and then delay for a couple of seconds). You will also see the angle printed out in the Shell.

Here's the code:

```python
from machine import Pin, PWM
from time import sleep

servo = PWM(Pin(32))
servo.freq(50) # pulse every 20ms

def set_angle(angle, min_pulse_us=500, max_pulse_us=2500):
    us_per_degree = (max_pulse_us - min_pulse_us) / 180
    pulse_us = us_per_degree * angle + min_pulse_us
    # duty 0 to 1023. At 50Hz, each duty_point is 20000/65535
    #     = 0.305 us/duty_point
    duty = int(pulse_us / 0.305)
    print(angle)
    servo.duty_u16(duty)

def waggle():
    set_angle(10)
    sleep(0.5)
    set_angle(90)
    sleep(0.5)
    set_angle(170)
    sleep(0.5)
    set_angle(90)
    sleep(0.5)

while True:
    waggle()
    sleep(2)
```

The PWM frequency is set to 50Hz (pulses per second) (`servo.freq(50)`), as that is the frequency of pulses that the servomotor expects.

Most of the work in generating the pulses to control the servomotor is contained in the function `set_angle`. This means that, if you want to create your own projects using servomotors, you can just copy this function.

The most important parameter for `set_angle` is the angle that you want to set the servomotor's arm to, which should be between 0 and 180. The other parameters set the minimum and maximum pulse width – you should not need to change these unless you have an unusual servomotor. The function starts by working out the number of microseconds of pulse which will be required for each degree of angle. It then calculates the total pulse length in microseconds for the angle required. Finally, it calculates the duty (the PWM value) between 0 and 65535, and uses `servo.duty_u16` to set that pulse width on the servomotor's control pin.

The `waggle` function sets the servo angle to 10 degrees, 90 degrees, 170 and then back to 90 degrees. If you are woindering why 10 and 170 were chosen rather than 0 and 180, it's because servomotors often have a truncated range, and reducing the angle range like this can prevent damage that can occur to the servomotor if it exceeds its range.

If you find that the servomotor judders badly at one end of its travel, you may need to adjust `min_angle` or `max_angle`.

The eternal loop just keeps calling `waggle`.

Battery Power

While you are working on a project, it's very convenient to power your ESP board from your computer's USB port. When it comes to cutting the umbilical cord and powering your projects independently of a computer, there are several options.

Perhaps the easiest way, is to continue to power the ESP32 board through its USB connection, by using a USB power adapter, such as a smartphone charger. These will generally be capable of providing far more power than the ESP32 needs (generally it needs about 0.5W) – but that's fine, as the board will just take what power it needs.

For battery power, you can use a USB backup battery, designed for providing extra battery power to your USB devices, however, because the ESP32 uses so little power, there is a danger that the USB battery pack decides that nothing is connected and turns itself off. So you may have to try a few different ones before you get a solution that stays on all the time.

If you are using an ESP32 Lite, then you will have noticed that it has a JST battery connector. The ESP32 Lite includes a LiPo (Lithium Polymer) battery charger on the board. This means that, if you buy a standard single cell (3.7V) LiPo battery with a JST lead, you can plug this into the ESP Lite (see Figure 8.14) and, whenever the board is connected to USB power, the battery will charge. When you unplug the USB lead, the board will immediately switch to battery power.

A typical battery such as this 420mAh one from Adafruit (`https://www.adafruit.com/product/4236`) can supply 420mA for an hour. Given that an ESP32 uses a maximum about 100mA, that should easily give 4 or more hours of battery operation.

Figure 8.14.   Powering an ESP32 Lite by LiPo battery.

Summary

In this chapter, we have explored the use of solderless breadboard to make some basic electronic circuits. In the next chapter, we will look at some more breadboard experiments using sensors.

CHAPTER 9

Sensors

In Chapter 8 we were mostly concerned with what we could call output devices such as LEDs and servomotors. In this chapter, we will look at sensors – devices that measure some physical property, such as the position of a knob, the temperature or the light level.

Analog sensors turn physical values such as light level, temperature or position into a voltage, that can be measured by an analog input of the ESP32. The experiments that we do in this chapter will, yet again, require a breadboard and some electronic components.

Variable Resistors

Variable resistors are, as the name suggests, resistors whose resistance can be varied. In this case, the resistance is varied by changing the position of a knob. Figure 9.1 shows such a variable resistor.

For historical reasons, relating to a common use of variable resistors to measure voltage, variable resistors are also often called *potentiometers* or just *pots*.

The resistor is made of a conductive track in an arc. The resistance between one end of the track and the other remains the same, but a central sliding contact (the *slider*) is moved by the variable resistor's knob, so that the resistance between the slider and one end of the variable resistor changes as you rotate the knob.

Resistance cannot be directly measured by an analog input. Analog inputs measure voltage, so we need a way of producing a variable voltage as the knob is turned. This is accomplished by the arrangement shown in Figure 9.2, called a voltage divider.

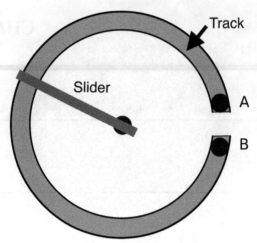

Figure 9.1. How a variable resistor works.

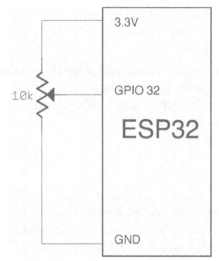

Figure 9.2. Using a variable resistor as a voltage divider.

One end of the variable resistor is connected to GND and the other to 3.3V, and the slider of the variable resistor is connected to an analog input (GPIO 32).

When the slider is at the GND end, the voltage at the slider will be 0V. When the slider is at the other end then it will be 3.3V and, at positions in-between, the voltage will be between 0 and 3.3V.

To test this out, wire up a variable resistor as shown in Figure 9.3, where the slider is connected to the centre pin of the pot. Note that on the ESP32 DevKit

1 GPIO 32 is sometimes labeled just *RX*.

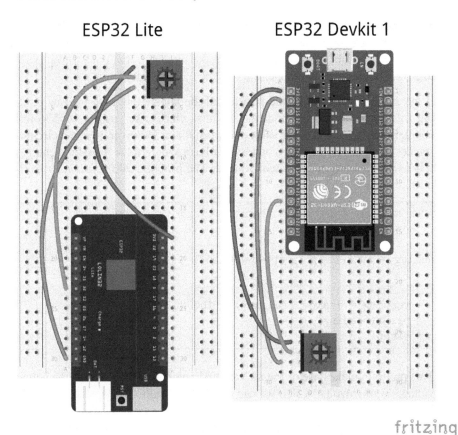

Figure 9.3. Breadboard Layouts for a Variable Resistor and ESP32 board.

To test the circuit, use the program *09_01_pot.py*.

Turn the knob on the variable resistor to see how the readings change between 0 and 65535.

```
from machine import ADC, Pin
from time import sleep

analog = ADC(Pin(32), atten=ADC.ATTN_11DB)

while True:
    reading = analog.read_u16()
    print(reading)
    sleep(0.5)
```

The code is much the same as that used in Chapter 7 when analog inputs were first introduced. The only difference is the additional parameter `atten=ADC.ATTN_11DB` when defining the ADC. This changes the upper voltage that the analog input reads from the default of 1.1V to about 3.1V. This means there will be a small dead-zone at the top end of the pot's travel (3.1V to 3.3V), but most of the range of 0V to the 3.3V supply is covered.

If we are just interested in the position of the pot's knob, then the raw reading values are fine. 0 will indicate the most clock-wise position, 65536 full counter-clockwise and other values between them. However, in some circumstances (perhaps monitoring a battery's voltage to determine its state of charge) it is useful to measure the actual voltage. The program *09_02_voltmeter.py* illustrates this.

```python
from machine import ADC, Pin
from time import sleep

analog = ADC(Pin(32), atten=ADC.ATTN_11DB)

while True:
    reading = analog.read_uv() / 1000000
    print(reading)
    sleep(0.5)
```

This program makes use of the ADC method `read_uv` which returns the actual voltage in microvolts, which we can divide by a million to get the voltage in Volts. The function `read_uv` is very convenient for measuring the voltage, but it still can't do anything about the dead-zones for the first 100mV and the last 200mV that will return minimum (0) and maximum readings respectively.

Sensing Light

You can use an ambient light sensor component such as the one included in the MonkMakes Basic Breadboard Kit. This component is actually a phototransistor – a component that acts a bit like a variable resistor, whose resistance varies depending on the amount of light falling on it.

Figure 9.4 shows the schematic diagram for connecting this sensor to an analog input.

The 1kΩ resistor converts the change in current (as more light hits the phototransistor) into a voltage, that can then be measured by the analog input. Wire up your breadboard as shown in Figure 9.5 to try this out. The longer lead of the phototransistor should be connected to row 3.

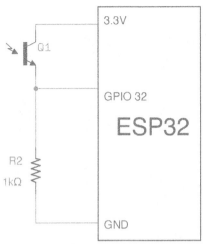

Figure 9.4. Schematic diagram for using a phototransistor to measure light intensity.

You can just use *09_01_pot.py* and *00_02_voltmeter.py* to get a reading for the light intensity. Or use *09_03_light_meter.py* which will measure the light intensity and display it as a percentage from 0 (totally dark) to 100 (bright sunlight). Run it and try shading the phototransistor with your hand – the light level should change.

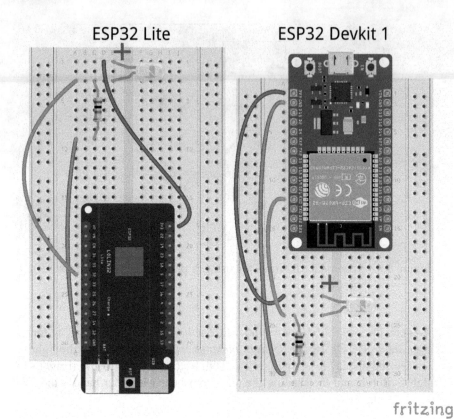

Figure 9.5. Breadboard Layouts for using a Phototransistor to Measure Light Intensity.

```
from machine import ADC, Pin
from time import sleep

analog = ADC(Pin(32), atten=ADC.ATTN_11DB)

max_reading = 58000

while True:
    reading = analog.read_u16()
    percent = reading / max_reading * 100
    print(percent)
    sleep(0.5)
```

The percentage is calculated using the proportion of the reading just taken to the maximum reading under the brightest conditions we are interested in. You can find this using *09_01_pot.py*.

The percentage reading is then clipped so that it is within the range 0 to 100.

Sensing Temperature

.

The TMP36 temperature sensor IC is a popular, low-cost analog temperature sensor. It is not the most accurate of devices (+/- 2 degrees Celsius) and if you need a more accurate measurement of temperature you might want to investigate the DS18B20 digital sensor.

Figure 9.6 shows the schematic diagram for using a TMP36 with an ESP32 board.

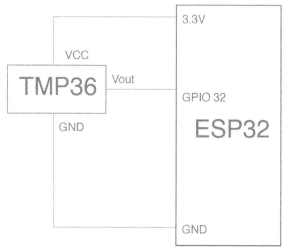

Figure 9.6.   Connecting an analog temperature sensor to an ESP32.

As well as power (*VCC* means +V supply) the TMP36 has a *Vout* pin to be connected to a GPIO pin acting as an analog input on the ESP32 board.

The breadboard layouts for this is shown in Figure 9.7. Take care to place the TMP36 the correct way around – it has a flat edge which should be to the left of the breadboard.

The program for this (*09_04_thermometer.py*) is listed below, and displays the temperature in degrees every half second.

ESP32 Lite   ESP32 Devkit 1

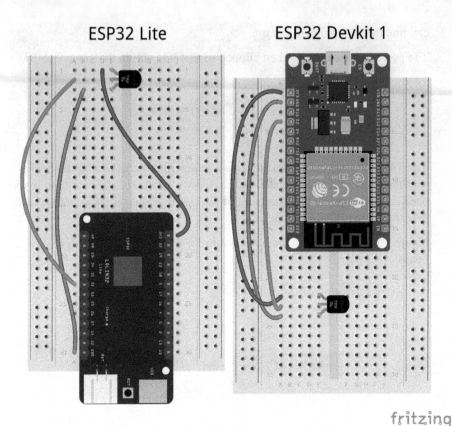

fritzing

Figure 9.7. Breadboard layout for using a TMP36.

```
from machine import ADC, Pin
from time import sleep

analog = ADC(Pin(32), atten=ADC.ATTN_11DB)

mv_per_c = 10
mv_offset = 500

while True:
    mv = analog.read_uv() / 1000
    temp_c = (mv - mv_offset) / mv_per_c
    print(temp_c)
    sleep(0.5)
```

From the spec sheet for the TMP36, we know that the relation between temperature and voltage is linear. In fact, each degree C results in an increase in voltage of 10 millivolts – so we set `mv_per_c` to 10 - and a temperature of 0 degrees C corresponds to a reading of 500 millivolts – so we set `mv_offset` to 500. If you are using a different analog temperature sensor with a different sensitivity and offset, it will be easy to change the code.

To convert the temperature into degrees Fahrenheit, you need to multiply the temperature in degrees C by 9/5 and add 32. The program *09_05_thermometer_f.py* does this.

Summary

In this chapter we have looked at various different sensors that can be connected to the analog inputs of the ESP32 board. In the next chapter, we will turn our attention to the ESP32's WiFi capabilities.

CHAPTER 10

WiFi

The ESP32 has the capability to connect to WiFi networks, act as a mini web server, or perform web requests to fetch data from internet web services.

The ESP32 is also technically capable of using Bluetooth. However, at the time of writing, the MicroPython bluetooth module does not support device pairing. Which is a shame, because you really need pairing to connect the ESP32 to a phone, or really do much of anything. So use of Bluetooth will have to wait for a later edition of this book, when Bluetooth is better supported.

Connecting to WiFi

Whatever you plan to do with your ESP32, the first step is likely to be getting it to connect to your home WiFi – just like any other electronic device. This will allow the ESP32 to communicate with other devices on your network and the internet.

> The ESP32's wireless hardware can only connect to 2.4GHz wireless networks. Most home networks operate at both 2.4GHz and 5GHz, often providing two access points – one 5GHz (not suitable for the ESP32) and another at 2.4GHz. This may be a feature that you have to access from your home router's admin console.

When you connect a device to your network, the network will allocate a unique name for your device within the network. This is called the devices *IP Address*. The code in *10_01_connect.py* will connect to a wireless network and then tell us the IP address that the network has allocated to the ESP32.

Before running it, change the following lines – replacing *network* with the name of your WiFi network and *password* with your WiFi password.

```
ssid = 'network'
password = 'password'
```

Now, when you run the program (if all is well), you should see the IP address that is allocated to your ESP32 displayed in the Shell (Figure 10.1).

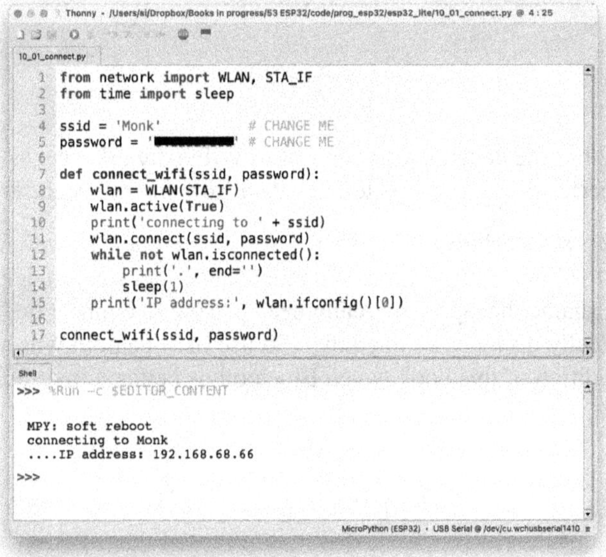

Figure 10.1. Connecting to a WiFi Network.

The real work is done in the function `connect_wifi` which takes two parameters – `ssid` (your WiFi network's name) and `password` (its password). The function creates an instance of the `WLAN` class with a parameter of `STA_IF` indicating that the ESP32 is a *station* (ordinary WiFi client). The program then ensures that the WiFi connection his active, and requests a connection using `wlan.connect(ssid, password)` if necessary. The `while` loop prints dots to indicate the progress in connecting, as this usually takes a few seconds. The dot printing `print` statement has an unusual second parameter of `end=' '` so that each dot isn't printed on a new line.

The `connect_wifi` function finds its way into all the code examples in this chapter.

Running a Web Server

Now that we can connect to a WiFi network, let's turn our ESP32 into a tiny web server. This means that the ESP32 will be able to serve a simple web page to a browser, connecting to it from any computer or other device on your network.

To make a web server you are going to download a module using Thonny. So, open the Thonny package manager (*Tools* menu then *Manage Packages..*). Search for *microdot*, select the matching package and click on *Install* to install it onto your ESP32.

Note. The Package manager is only available if your ESP32 is connected to your computer.

> As well as the standard MicroPython modules, there are also many contributed modules written by people like me and you and contributed to the MicroPython community for free. When a module is made available to the public, it is usually installed by wrapping it up as a *package* and using the *pip* (Pip Installs Packages) tool. pip will download and install MicroPython (and regular Python) packages from the *PyPi* website.

Thonny makes this service very easy to use, by providing a *Package Manager* available from the *Manage packages* option from the *Tools* menu (Figure 10.2).

Figure 10.2. The Thonny Package Manager.

You can type the name of the package you are looking for into the search field, find it and then install the module. When you install a package containing a module in this way, it is copied onto the file system of the ESP32, so that it is available for your code to use.

Once installed, open the program *10_02_hello_web_server.py*. Again, you will need to change `network` and `password` to suit your network.

If the server is running correctly, you should see the IP address in the Shell (Figure 10.3) as before. Copy this address onto the clipboard, and then paste it into a browser window on your computer - you should see should see something like Figure 10.4.

Figure 10.3.   An ESP32 Web Server.

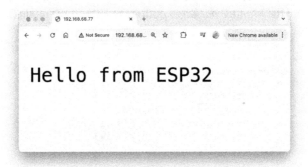

Figure 10.4.   Viewing a Web Page Running on a ESP32 (zoomed in).

Let's take a look at the code for this. For brevity, the `connect_wifi` function is truncated.

```
from network import WLAN, STA_IF
from microdot import Microdot
from time import sleep

ssid = 'network'
```

```
password = 'password'

def connect_wifi(ssid, password):
...

app = Microdot()
connect_wifi(ssid, password)

@app.route('/')
def index(request):
    return 'Hello from ESP32'

app.run(port=80)
```

What's new is the use of the microdot module. A new instance of the class Microdot is created and assigned to the variable app.

The way that the microdot framework works is that, when a request comes to the web server from a browser, a function is made responsible for *serving* the page being requested. In this case, that function is index. Using @app.route('/') line before the function tells microdot that this is the handler for the root page (/).

The index function simply returns the string 'Hello from ESP32', which the microdot framework will send to the browser requesting the page.

The line app.run(port=80) starts the web server listening for requests for browsers on port 80 (the default web server port).

A WiFi Lightmeter

We can build on our web server example to have the ESP32 serve a more interesting page. Let's extend our light-measurement example, and serve its results as a nice-looking graphical gauge as shown in Figure 10.5.

The project assumes that you have the made the light-sensor example (from Chapter 9 – See page 107) attached to the ESP32. If you do not have any hardware connected to the ESP32, this example will still display analog readings, but they will display as random values.

Before we get to the graphical gauge display, let's modify *10_02_hello_web_server.py* so that instead of just displaying a message, we will display the light level. You can find the code for this in *10_03_web_lightmeter.py*. This is listed below, with the connect_wifi truncated.

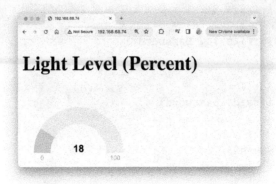

Figure 10.5. A WiFi Lightmeter using a Gauge Control.

```python
from network import WLAN, STA_IF
from microdot import Microdot
from machine import ADC, Pin
from time import sleep

ssid = 'network'
password = 'password'

max_reading = 58000

sensor = ADC(Pin(32), atten=ADC.ATTN_11DB)

def connect_wifi(ssid, password):
    ...

def read_light():
    reading = sensor.read_u16()
    percent = int(reading / max_reading * 100)
    if percent > 100:
        percent = 100
    return percent

app = Microdot()
connect_wifi(ssid, password)

@app.route('/')
def index(request):
    return 'Light: ' + str(read_light())

app.run(port=80)
```

This program is a mixture of *10_02_hello_web_server.py* and *09_03_lightmeter.py* from Chapter 9. The `index` function uses `read_light` to read the light value from an analog input, and append it to the string `'Light: '`, before returning the result to the browser calling it.

When you run it, you should see something like Figure 10.6. To get the light level to update, you will need to refresh the page.

Figure 10.6. Viewing a Lightmeter Web Page Running on a ESP32.

Now that we have the basic lightmeter, let's improve it by using some fancy web technology to display the attractive gauge shown in Figure 10.5. As you might expect, the code for this is a lot more complex, and can be found in the file *10_04_web_lightmeter_gauge.py*

Most of the added complexity of this example takes place in the web browser, using a programming language called JavaScript. This code is contained in the `index_page` variable's multi-line string shown below. As the name suggests, *multi-line* strings are strings that span multiple lines. In Python, they are marked by a triple quote at the start and the end.

```
index_page = '''
<!DOCTYPE html>
<html>
<head>
  <script src="http://ajax.googleapis.com/ajax/libs/jquery/1.7.2/jquery.min.js" type="text/javascript"
          charset="utf-8"></script>
  <script src="http://cdnjs.cloudflare.com/ajax/libs/raphael/2.1.0/raphael-min.js"></script>
  <script src="https://cdnjs.cloudflare.com/ajax/libs/justgage/1.6.1/justgage.min.js">
  </script>
  <script>
  function callback(lightStr, status){
    if (status == "success") {
      light = parseFloat(lightStr).toFixed(2);
      g.refresh(light);
      setTimeout(getReading, 1000);
    }
    else {
      alert("There was a problem");
    }
  }
  function getReading(){
    $.get('/light', callback);
  }
  </script>
```

```
</head>

<body>
<h1>Light Level (Percent)</h1>
<div id="gauge" class="200x160px"></div>

<script>
var g = new JustGage({
    id: "gauge",
    value: 0,
    min: 0,
    max: 100,
});
getReading();
</script>
</body>
</html>
'''
```

Just to emphasise, this code does not run on the ESP32. It is just passed from the ESP32 to any web browser requesting the root index page. For reasons of space we can't get too much into JavaScript, but we can go through roughly how the code works.

Since we are sending back a webpage to display in a browser, it must have certain HTML (HyperText Markup Language) code such as the <html>, <head> and <body> tags. At the start of the the <head> tag there are three script tags. These fetch JavaScript libraries (like modules), that the web page needs, from the internet.

After this, there is another script tag that contains two JavaScript functions – callback and getReading. The getReading function makes a web request to the \light web page served by the ESP32. The callback function will be run when the web request returns a value of light reading from the ESP32, and then sets a timer to run another web request in a second (1000 milliseconds).

Inside the body tag, the <h1> tag specifies that Light Level (Percent) should be displayed as a level 1 heading. The <div> tag then reserves an area, with an id of "gauge", in which the gauge should be displayed.

Next, we have another <script> tag, containing some more JavaScript code, that creates a gauge g and then calls getReading to start the display refreshing.

The rest of *10_04_web_lightmeter_gauge.py* is very similar to the previous example, except that this time, as well as /index (which sends the index_page string to the browser) there is now a second page, called /light. The /light page converts the light reading to a string and then sends it back to the JavaScript function getReading that requested it from the browser.

```
app = Microdot()
connect_wifi(ssid, password)

@app.route('/')
def index(request):
```

```
    return index_page, 400, {'Content-Type': 'text/html'}

@app.route('/light')
def temp(request):
    return str(read_light())

app.run(port=80)
```

Calling a Web Service

All the examples, so far have involved running a local web server on the ESP32. In this next example, we will look at how the ESP32 can act like a browser, and use data that it fetches from the internet to do something – in this case, to use the *open weather map* web service to lookup the current outdoor temperature and weather description, for a particular location, and display this information in the Shell (Figure 10.7).

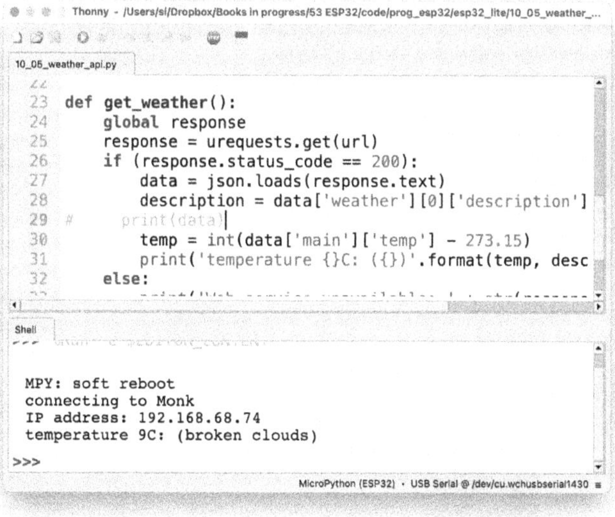

Figure 10.7. Calling a Weather Web Service API.

The Open Weather Maps service is free if you keep your number of requests to their API (Application Programming Interface) below one million calls a month; to use it, you will need to register for their free tier at: https://openweathermap.org/. Registering will give you an access key to allow you to connect to their service to get their weather data. To get your key, log in, then click on the dropdown menu on your account name and select the option *My API Keys* (Figure 10.8).

Figure 10.8. Creating an Open Weather Maps Key.

Create a new key by entering a name (anything will do for a name) and then clicking the *Generate* button. This creates a long key, that you will need to copy and paste into the Python program.

You also need to specify a latitude and longitude for the weather API. One way to find this information is to open Google maps in your browser and pick a location. When you click on it, the latitude and longitude will pop-up (Figure 10.9).

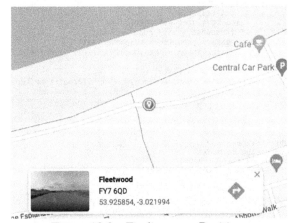

Figure 10.9. Finding your Position.

The code is quite long, so open the program *10_05_weather_api.py* while its operation is described; there are also the usual changes you will need to make to the program for your network credentials and the key you have just obtained.

We create a block constants associated with the program, for things like your WiFi credentials and the key. You will need to edit these, before the program will work.

```
ssid = 'network'
password = 'password'
key = 'ea751fc7712f27059e8agh99445453b712'
location = 'lat=53.925854&lon=-3.021994'
api_base = 'http://api.openweathermap.org/data/2.5/weather?'
url = api_base + location + '&appid=' + key
```

Change the values of ssid and password to match those of your WiFi network and then paste in new values for lat, long and key with the key that you generated earlier on the open weather website.

The full URL (url) for the call to the weather API is created by concatenating the variables you have just been editing.

After this, we have our usual WiFi connection function and then the function get_weather, where most of the action takes place.

```
def get_weather():
    response = urequests.get(url)
    if (response.status_code == 200):
        data = json.loads(response.text)
        # print(data)
        description = data['weather'][0]['description']
        temp = int(data['main']['temp'] - 273.15)
        print(f'temperature {temp}C: ({description})')
    else:
        print('Web service unavailable: ' +
            str(response.reason, 'utf-8'))
```

The get_weather function uses the urequests module to call the web service using the url constructed earlier. If the call is successful, then the API will respond with a status_code of 200 and the data that we asked for.

This data it returns will look something like this, which you can see by removing the # sign in front of print(data).

{'timezone': 3600, 'sys': {'type': 1, 'sunrise': 1697438439, 'country': 'GB', 'id': 1421, 'sunset': 1697476469}, 'base': 'stations', 'main': {'temp_min': 282.2, 'pressure': 1020, 'feels_like': 281.41, 'humidity': 71, 'temp_max': 284.2, 'temp': 282.98}, 'visibility': 10000, 'id': 2649312, 'clouds': {'all': 0}, 'coord': {'lon': -3.022, 'lat': 53.9259}, 'name': 'Fleetwood', 'cod': 200, 'weather': [{'id': 800, 'icon': '01d', 'main': 'Clear', 'description': 'clear sky'}], 'dt': 1697475873, 'wind': {'speed': 3.09, 'deg': 110}}

This response, from the API (in the variable data) is in a format called *JSON* (JavaScript Standard Object Notation). If you look closely at it, it is made up of attribute and value pairs grouped together in curly braces (just like a Python dictionary) and separated by commas. In the case of 'weather' this value is contained in square brackets (like a Python list). But, in this case, the list just has one entry which also contains the equivalent of a Python dictionary. We will need to extract the temperature (`temp`) and description (`description`) values from this JSON.

To extract the `description` we use `data['weather'][0]['description']`. In other words, the `description` contained within the first element (there is only one) of the `weather` list.

Extracting the temperature is a bit simpler, as this is just available as the `temp` attribute of the `main` dictionary, so we use `['main']['temp']`. The temperature is in degrees Kelvin, so to convert this to Celcius, we need to subtract 273.15. `get_weather` then prints both the temperature and the description.

If you have problems with the web services request working, you can test it out in a browser tab by pasting the URL into the address bar. The URL will just be in the variable `url` that you can find by pressing CTRL-c while the program is running and then entering `url`.

```
>>> url
'http://api.openweathermap.org/data/2.5/weather?lat=
53.925854&lon=-3.021994&appid=ea751fc7712f27059e8
agh99445453b712'
>>>
```

This project could easily be adapted to display other information about the current weather, or even display information from other APIs provided by Open Weather Maps, such as weather forecast information.

Summary

In this chapter, we have seen how to use a ESP32 as a tiny web server displaying light readings and also calling a web service to retrieve weather data. Both projects should give you a basis for your own internet ESP32 projects.

CHAPTER 11

Displays

In this chapter, we will look at how to connect and use two common types of display with the ESP32 board. The first is a common graphical OLED display. These displays are low-cost and readily available and, although a little small, are popular and easy to use.

The second type of display uses addressable LEDs. These LEDs, often called NeoPixels, can be found in various arrangements, from long lengths of regularly spaced LEDs on tape, to rectangular arrays of LEDs.

OLED Displays

Low-cost OLED (Organic LED) displays, like the one shown in Figure 11.1, are widely available on the internet. Search for SSD1306 OLED, and look for displays that are 128x32 or 128x64 pixels in size. *SSD1306* is the driver chip used by the OLED display.

These displays are bright and clear, and are a stylish way of getting you ESP32 board to display text or graphics.

Hardware

Figure 11.2 shows how to connect an OLED display to your ESP32 board. The OLED display requires 3V power, and the SDA and SCL pins (on some OLED boards SCL is labeled SCK) connections of the display are connected to GPIO 19 and 18 respectively, in what is known as a I2C interface. The ESP32 Devkit 1 does not have a pin 18, and so pin 21 is used.

Figure 11.1. A low-cost 128 by 64 pixel OLD display and ESP32 board on breadboard.

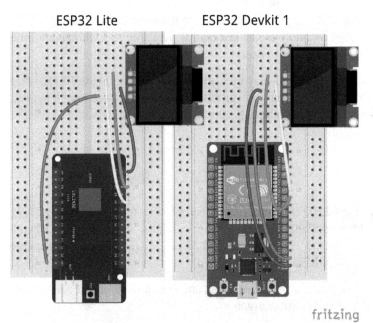

Figure 11.2. Wiring an OLED display to a ESP32 board.

Note that some OLED displays have the pins in a different order, so check what is written on the OLED display itself next to the pins.

The connections are:

- GND of the ESP32 board to GND of the OLED display

- 3V of the ESP32 board to VCC of the OLED display

- 19 of the ESP32 board to SDA of the OLED display

- 18 (or 21 for Devkit 1) of the ESP32 board to SCL of the OLED display

The I2C interface usually requires two pull-up resistors to be used; however, the ESP32 board's I2C interface can use the ESP32's internal pull-up resistors, so external resistors are not required.

Software

Using an OLED display like this requires the use of a Python module. Open Thonny's package manager (*Tools* menu then *Manage Packages....* Type *SSD1306* in the search area and then click *Search PyPI'*. The top result should be *micropython-ssd1306*. Select this and install it.

Now that the module is on the ESP32 board's file system, any program that we upload that needs to use the module will have access to it. So, open the program *11_01_oled.py* and run it. Of course, if you want it to start automatically, you will have to save a copy, renaming it as *main.py*, on the ESP32 board.

Here's the code for the project that displays a text message with a rectangular border, as shown in Figure 11.3.

Figure 11.3. Showing a Message on an OLED display.

```
from machine import Pin, I2C
from ssd1306 import SSD1306_I2C

i2c = I2C(0, sda=Pin(19, pull=Pin.PULL_UP),
          scl=Pin(18, pull=Pin.PULL_UP))

oled = SSD1306_I2C(128, 64, i2c)

oled.fill(0)
oled.rect(0, 0, 127, 63, 1)
```

```
oled.text('Programming the', 5, 12, 1)
oled.text('ESP32', 5, 22, 1)
oled.text('by', 5, 32, 1)
oled.text('Simon Monk', 5, 42, 1)

oled.show()
```

From the machine module, we need to import Pin and also I2C. I2C (pronounced *I squared C*) is a standard way of connecting displays and other modules to a microcontroller like the ESP32. It uses two GPIO pins, and sends data back and forth at high speed.

A variable i2c is used to represent the I2C interface that we are going to use on the ESP32 board. The ESP32 board actually has two I2C channels and each of these channels can be allocated to different pair of pins. We are going to use the first I2C channel (0) on pins 19 and 18. We specify this by using the following line, which also turns on the pull-up resistors:

```
i2c = I2C(0, sda=Pin(19, pull=Pin.PULL_UP),
          scl=Pin(18, pull=Pin.PULL_UP))
```

The parameter sda refers to the I2C data connection, and scl to the clock connection.

The next line declares a variable oled to represent the display. The first two parameters are the display resolution (in this case 128 wide by 64 high); the final parameter links the oled display to the i2c interface.

Now that the display is all set up, we can display text and graphics on it. The display used here is monochrome (blue on black); other colors are available, and also multicolor oled displays. However, for this display, a value of 0 signifies a pixel is off and 1 that it is on. So, to clear the entire screen, we can fill it with 0 like this:

```
oled.fill(0)
```

To draw a rectangle we need to specify the starting x and y coordinates (0, 0 is top left) and then the width and height of the rectangle in pixels (in this case 127 x 63). The final parameter to oled.rect is the color (in this case 1).

We can also display text, in which case the parameters are the text to display, followed by the x and y coordinates for the top left of the text, followed by the color (1).

All of these drawing commands will have no effect on the display until we call `oled.show`. It us usual to prepare all of the things you want to display first, and then use a single `oled.show` – rather than do `oled.show` after every drawing command.

When you run the program, you may see this error:

```
>>> %Run -c $EDITOR_CONTENT
Traceback (most recent call last):
  File "<stdin>", line 6, in <module>
  File "/lib/ssd1306.py", line 110, in __init__
  File "/lib/ssd1306.py", line 36, in __init__
  File "/lib/ssd1306.py", line 71, in init_display
  File "/lib/ssd1306.py", line 115, in write_cmd
OSError: [Errno 5] EIO
>>>
```

This indicates that the ESP32 board is unable to establish communication with the display. The most common reason for this is that it just isn't wired up correctly, so check your wiring. The other likely cause of this error is that the display has a different I2C address from the default (3C in hexadecimal or 60 in decimal).

You can find the address of your display by running the following command in the Shell, after running the program. There is no need to stop the program, using the stop button or CTRL-c, because the program will have stopped by itself – there is no eternal loop.

```
>>> i2c.scan()
[61]
>>>
```

This command returns a list of attached I2C devices (we are only expecting one). So, in this case, we can see that the address is 61, not the default of 60. We can fix our program by adding a new `addr` parameter to the end of line 4.

```
i2c = I2C(0, sda=Pin(19, pull=Pin.PULL_UP),
          scl=Pin(18, pull=Pin.PULL_UP), addr=61)
```

If the `i2c.scan()` returns an empty list, then there is either a problem with the wiring, or the OLED display is broken.

An OLED Clock

This project uses the same breadboard layout as Figure 11.2 and you will find the program in *11_02_oled_clock.py*.

When you run the program, the time will be displayed on a conventional look-ing clock face, alongside the current date. This first version of the program uses the time that the ESP32 board thinks it is. This time is, by default, au-tomatically set when a program is run in Thonny. Note that for this to work, the two options shown in the Thonny Interpreter options (Figure 11.4) must be checked.

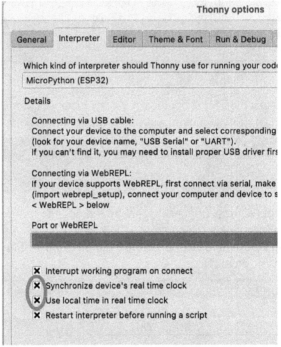

Figure 11.4. Time Sync Settings in Thonny.

The ESP32 will keep the correct time until it is powered down. Without another way of setting the time, this does not make a very useful clock if the project is going to run stand-alone, perhaps on battery power. In a later second version of the program, we will make use of the ESP32's WiFi capabilities to fetch the current time from a network time server on the Internet.

Here's the first version of the clock (*11_02_oled_clock.py*).

Figure 11.5. An OLED analog clock.

```python
from machine import Pin, I2C, RTC
from ssd1306 import SSD1306_I2C
from math import sin, cos, radians
from time import sleep, localtime

i2c = I2C(0, sda=Pin(19, pull=Pin.PULL_UP),
          scl=Pin(18, pull=Pin.PULL_UP))
oled = SSD1306_I2C(128, 64, i2c)

origin_x = 32
origin_y = 31

months = ['Jan', 'Feb', 'Mar', 'Apr', 'May', 'Jun',
          'Jul', 'Aug', 'Sep', 'Oct', 'Nov', 'Dec']

def show_time():
    oled.fill(0)
    draw_face()
    dt = localtime()
    hour, minute, second = dt[4], dt[5], dt[6]
    draw_hand((hour + minute / 60) * 5, 16)
    draw_hand(minute + second / 60, 20)
    draw_hand(second, 25)
    draw_date(dt, 70, 10)
    oled.show()

def draw_date(dt, x, y):
    day, month, year = dt[2], dt[1], dt[0]
    date_str = str(day) + ' ' + months[month-1]
    year_str = str(year)
```

```
    oled.text(date_str, x, y, 1)
    oled.text(year_str, x, y+40, 1)

def polar_rect(angle, radius):
    x = int(radius * sin(radians(angle)))
    y = -int(radius * cos(radians(angle)))
    return x, y

def draw_hand(minutes, radius):
    angle = (360 / 60) * minutes
    x, y = polar_rect(angle, radius)
    oled.line(origin_x, origin_y, origin_x+x, origin_y+y, 1)

def draw_face():
    for hour in range(0, 12):
        x, y = polar_rect(hour * 30, 28)
        hour_str = str(hour)
        if hour == 0:
            hour_str = '12'
        if hour > 9 or hour == 0:
            x -= 3
        oled.text(hour_str, origin_x+x-4, origin_y+y-3, 1)

while True:
    show_time()
    sleep(0.1)
```

There is quite a lot going on in this program, including a bit of trigonometry to display the clock hands and face.

The variables origin_x and origin_y define the center point for the clock face – in this case, centered vertically on the left-hand side of the OLED display.

The months array will be used later in the code to convert a month number into the more user-friendly abbreviated month name.

Next we have a series of functions used to draw the screen, starting with the function show_time. This clears the screen and then calls draw_face, which we will come to later. It then fetches the current hours, minutes and seconds, and draws each of the hands using draw_hand. Let's now look at draw_face and draw_hand.

These functions rely on something called *polar* coordinates, if you have not met this concept before, here's how it works. Figure 11.6 shows how you can specify the position of a point in one of two different ways.

The way we are used to is to specify the x and y coordinates of the point,

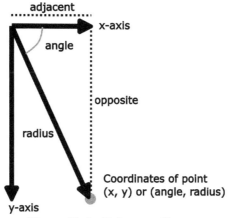

Figure 11.6. Polar coordinates.

relative to the origin at x=0 and y=0, which in the case of an OLED screen is at the top left.

An alternative way of referencing the location of the point is to use polar co-ordinates, that involves specifying an *angle* and a distance or *radius*. Using polar coordinates works much better for drawing a clock face or lines for clock hands, because its easy to convert say the hour position on a clock (1 to 12) into an angle (0 to 360) and the length of the hand is just indicated by the radius. The function `polar_rect` does this in our code.

```
def polar_rect(angle, radius):
    x = int(radius * sin(radians(angle)))
    y = -int(radius * cos(radians(angle)))
    return x, y
```

The function takes an `angle` and `radius` as parameters, and returns x and y coordinates. The x coordinate is found using the `sin` function. The sin of an angle is the ratio of the lengths of the opposite side (opposite the angle) to the hypotenuse (the radius). The y coordinate is calculated using the `cos` (cosine) function, which is the ratio of the adjacent to the hypotenuse (radius).

The function `draw_hand` takes a parameter `minutes` (0 to 59). Since there are 360 degrees in a full circle and 60 minutes in a circle, each minute has an angle of 360/60 degrees. This is then multiplied by the number of `minutes` to give the total angle for the hand. The length of the hand is the second parameter `radius`. This is converted into x and y coordinates using `polar_rect` that is then used to draw a line from the origin to x, y.

Drawing a clock face using `draw_face` involves counting through the hours, and then writing text for the hour at a radius of 28 pixels around the clock face. There are some complications, in that if the hour is 0, then 12 should appear on the clock face. Also, if the hour is double digit, we need to offset the text a little to center it in the right position on the clock face. A degree of trial and error was involved in getting this right.

A Self-Setting OLED Clock

Since the ESP32 can talk to the Internet, we can make sure that it always has the correct time by calling a network time server once a day.

To do this, we need to use the Package Manager to install a package called *datetime* onto the ESP32. So, as you did with the ssd1306 package, open the Package Manager and search for *datetime* and install it.

The code for this improved program can be found in *11_03_oled_clock_ntp.py*. Much of it is the same as *11_02_oled_clock.py* so we will just highlight what's new.

When you get the time from a time server using the standard NTP (Network Time Protocol), the time will be in UTC. That is, there is no consideration for which time-zone you are in. So, we can use the timezone function from the datetime package to set the timezone like this:

```
tz = datetime.timezone(datetime.timedelta(hours=0))
```

Here hours is set to 0 for no time difference; if you are on BST in the UK you would set hours=1, and if you are on EST in the US you would set hours=-5.

When we want to get the correct time, we can pass the timezone as a parameter to the now method of datetime module like this:

```
datetime.datetime.now(tz)
```

This will return a datetime object adjusted to the local time zone.

The function set_time is listed below. It first connects to WiFi and then calls settime in the ntptime module to update the ESP32's time.

```
def set_time():
    connect_wifi(ssid, password)
    message('UPDATING ....')
    try:
        ntptime.settime()
        message('Update Success', delay=2)
```

```
except:
    message('NTP fail', delay=2)
oled.show()
```

Instead of `print` commands to monitor the process of connecting and getting the time from the internet, we use the function `message` that, more usefully, displays the message on the OLED screen.

To fetch the time every day at 1am in the morning, the eternal loop now looks like this:

```
while True:
    t = datetime.datetime.now(tz)
    if t.hour == 1 and t.minute == 0 and t.second == 0:
        set_time()
    show_time(t)
    sleep(0.1)
```

If the hour is 1 and the minute and seconds are both 0, then `set_time` is called. Fortunately, since `set_time` takes a few seconds to run, there is no chance of it being run multiple times on the same day.

NeoPixel Displays

NeoPixel has become a popular term for displays using chains of addressable LEDs – LED chips that include red, green and blue LEDs, as well as control logic to set the color and brightness. Such displays are available in all sorts of shapes and sizes, from long tapes with adhesive backing (Figure 11.8), or rings of LEDs on a PCB, to rectangular arrays of LEDs arranged as a screen with very large pixels.

Hardware

Adafruit have a nice range of NeoPixel displays, and you can also find similar products at a lower cost (made in China). The key thing to look for is WS2812, which is the name of the NeoPixel addressable LED chip used for the displays.

Theses displays require a logic level that is close to the supply voltage; by preference this would be 5V but, although the ESP32 board can supply 5V at a reasonable current, its logic level is 3.3V. So, if you try to supply the display with 5V, but use the 3V data logic of the ESP32 board, the results are unpredictable. Hence, it is better to use a 3V supply. This means that there is

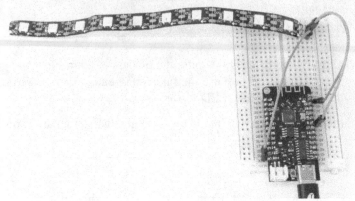

Figure 11.7. A NeoPixel display controlled by ESP32.

much less current available to power the LEDs, so you will need to limit either the brightness of the LEDs and/or the number of LEDs. Otherwise the display will take too much current, the voltage will then drop, and the display and/or ESP32 board will not operate reliably. So, if things start to misbehave, that's probably what is happening and you need to reduce the brightness of the LEDs or use a separate high current power supply to provide power to the NeoPixels.

However many LEDs you have in your display, you will have just three connections to be made—power (3V and GND) and data (which is connected to a GPIO pin). In the code example in the next section, it is assumed that the data connection is connected to pin GP22.

You can make the connections using breadboard, as shown in Figure 11.8, where wires are soldered to the display and then connected to the breadboard. Alternatively, you may find that you can connect to the display using jumper wires, as some LED strips come with sockets into which jumper wires can be fitted. Neopixel LED strips have connections at both ends. Make sure you connect to the input end, which is often marked with an arrow pointing into the strip.

Software

MicroPython includes a library for controlling NeoPixels, so you don't have to add any modules to your ESP32. Here's the code for a test program (*11_04_neopixel.py* that allows you control the LEDs in various ways, by entering commands at the Shell. Entering *c* will turn all the LEDs off, *r* will set all the LEDs to a different random color, and entering a number will set the LED at that position to white.

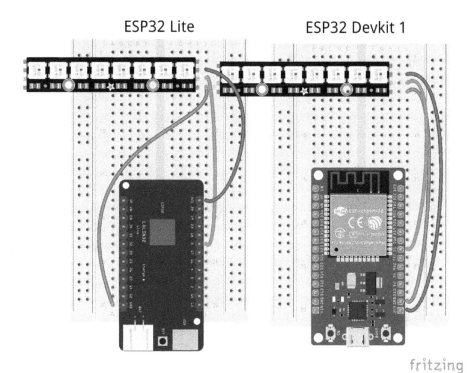

ESP32 Lite   ESP32 Devkit 1

fritzing

Figure 11.8. Breadboard Layouts for a NeoPixel display controlled by ESP32.

Before running the program change the value of NUM_LEDS to the number of LEDs on your display.

```python
from time import sleep
from machine import Pin
from random import randint
from neopixel import NeoPixel

NUM_LEDS = 10

pixels = NeoPixel(Pin(5), NUM_LEDS)

def clear():
    pixels.fill((0, 0, 0))
    pixels.write()

def randomize():
    clear()
    for i in range(NUM_LEDS):
        pixels[i] = (randint(0, 50), randint(0, 50),
```

```
        randint(0, 50))
    pixels.write()
    sleep(0.1)

randomize()

print("Enter the LED's number to turn it on")
print("or c-clear r-randomize")
while True:
    led_str = input("command: ")
    if (led_str == 'c'):
        clear()
    elif (led_str == 'r'):
        randomize()
    else:
        led = int(led_str)
        pixels[led] = (50, 50, 50) # white
        pixels.write()
```

An instance of NeoPixel is created using the line pixels = NeoPixel(Pin(5), 10). Two parameters are passed, the GPIO pin to be used to control the NeoPixels and the number of Neopixels in the strip.

The clear function uses the fill method to set all the LEDs to off (the color (0, 0, 0)) where each of the three numbers is a value between 0 and 255 representing the brightness of the LED in (R, G, B). The LED strip won't actually change colors until the write method is called.

To set the color of a particular LED, you treat pixels like an array and use the square bracket notation like this: pixels[led] = (50, 50, 50).

Summary

In this chapter we have used two very different types of display. There are, of course, many other types of display available, many of them using an I2C interface. If you have hardware that you want to connect to your ESP32 board, search to see if someone has written a module for it – this is a lot easier than trying to write your own.

CHAPTER 12

Advanced Input Output

In this chapter, we'll explore some more advanced features of the ESP32 board that you might find useful.

The ESP32 board's IO (Input Output) capabilities outstrip a lot of microcontrollers. These include high speed PWM available on most pins, true analog outputs and, perhaps most impressively, there are two cores (processors) in the ESP32, allowing the ESP32 to do two things at once.

Interrupts

So far, when we have needed to detect that a button has been pressed, we have put some code like this example from *08_03_RGB_switch.py* in the eternal loop:

```python
button = Pin(17, Pin.IN, Pin.PULL_UP)

while True:
    if button.value() == 0:
        # do something
        sleep(0.2) # debouncing
```

This way of doing things is called *polling*, because we just keep checking the state of the digital input over and over again until we read a 0, indicating a button press. This is fine for a simple program like this, where nothing else is happening, but imagine the situation where we were doing something time-consuming in the loop – perhaps reading a sensor value and displaying it. In that case, there would be a risk that a very quick button press might happen when the code was busy doing something else, and be missed.

For situations like this, you can use *interrupts*. A digital input can be specified as causing an interrupt. As the name suggests, when an interrupt is detected whatever code is currently being run is suspended and an interrupt service routine (ISR) is run. When the ISR has finished so, the program continues where it left off. So, we have not wasted any time polling for a button press.

Let's have a look at an example of this. This interrupt is triggered by the change in input voltage of a digital input – for example when a switch is pressed. To try this out, connect a switch between GPIO 14 and GND, using the breadboard layout of Figure 12.1. The switch neatly fits between GPIO 14 and GND on a ESP32 Lite.

If you don't have a push switch, then you can use GPIO 12, which is right next to GND and make the connection by touching a screwdriver or paper-clip between the pins as described in Chapter 7 (see page 78).

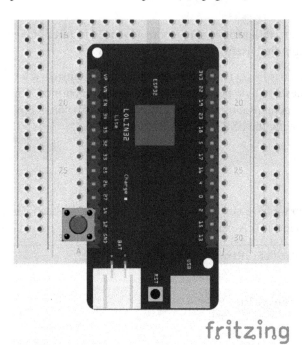

Figure 12.1. A switch to test interrupts.

Now run the program *12_01_interrupts.py*.

```
from machine import Pin
from time import sleep

button = Pin(14, Pin.IN, Pin.PULL_UP)

def handle_button(ignore):
```

```
    print('BUTTON PRESSED')

button.irq(handle_button, Pin.IRQ_FALLING)
i = 0

while True:
    i += 1
    print(i)
    sleep(0.2)
```

The code that we want to be run following an interrupt must be placed in a function. The `irq` (interrupt request) method on `button` associates the interrupt handler function (`handle_button` in this case) with the `button` pin.

Note that the `handle_button` takes a parameter that we have chosen to call `ignore`. This parameter will contain a reference to the pin that caused the interrupt. If you use a different interrupt handler function for each button, as in this example, then this can be ignored.

We want the interrupt to happen when the input pin button goes from high to low (i.e. when it is first pressed, rather than when it's released) and so the second parameter to `irq` is `Pin.IRQ_FALLING`. If you wanted the interrupt to happen when the button was released, rather than when it was pressed, then you would specify `IRQ_RISING`.

If your project requires more than one interrupt, that is fine – you can associate other pins with other interrupt handler functions.

Timer Interrupts

As well as catching interrupts from a digital input, you can also set periodic interrupts from a hardware timer, running on the ESP32 board. For example, we could change the LED blinking program to use a timer interrupt, so that it works without the need for calls to `sleep`. This frees up the eternal loop for other activities.

You can find this program in *12_02_blink_timer.py*.

```
from machine import Pin, Timer
from time import sleep

led = Pin(22, Pin.OUT)

led_state = 0
```

```
def tick(timer):
    global led, led_state
    led.value(led_state)
    led_state ^= 1

timer = Timer(0)

timer.init(freq=2, mode=Timer.PERIODIC, callback=tick)

x = 0
while True:
    print(x)
    x += 1
    sleep(1.2)
```

Just to make the program a bit more interesting, the eternal loop counts up in the Shell at the same time as the LED is flashing – demonstrating that the timer is not adversely affecting anything else going on in the program.

The `tick` function is associated with a timed interrupt using the `init` method on a `Timer` instance that is created by the line `timer = Timer(0)`. The ESP32 has four hardware timers numbered 0 to 3. Here we are using timer 0.

The frequency of interrupt (times per second) is specified in the `freq` parameter. The mode of `PERIODIC` means the interrupt will keep repeating. You can also specify `ONE_SHOT`, if you only want the timed interrupt to happen once. You can find out more about the `Timer` class at `https://tinyurl.com/cxupyupk`.

It is a good idea to keep the code in an interrupt service routine as short and quick as possible because, even though it may look like the ESP32 board is doing two things at the same time, the counting is actually interrupted for a very short amount of time while the LED is being toggled.

To toggle the LED, a global variable `led_state` is used. The code `led_state ^= 1` changes the variable to a 0 if its 1 and 1 if its 0. Rather than use the on and off methods of `Signal` we are turning the LED on and off using `led.value`.

High-Speed PWM

The ESP32 is not unusual in providing a PWM output – almost all modern microcontrollers do this but, what is unusual is the resolution and speed of the PWM that the ESP32 board can produce. If you are controlling the brightness of an LED, or the speed of a motor, then you do not need very high speed

PWM. However, if you are producing audio signals, then high speed and high resolution make it much easier to produce high-quality analog output signals.

The ESP32 board is theoretically capable of producing a 16 bit PWM output at a frequency of 40MHz. See Figure 12.2 where program *12_fast_pwm.py* is being used to turn a GPIO 32 on and off a million times a second. Here the signal has been sent to an output pin, and displayed on an oscilloscope rather than an LED, since the human eye cannot distinguish anything flashing this fast.

```
from machine import Pin, PWM

out_pin = PWM(Pin(32))
out_pin.freq(1000000) # 1MHz

out_pin.duty_u16(32000) # 50%
```

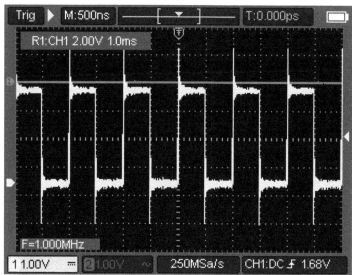

Figure 12.2. High-speed PWM – 1MHz, 50% duty cycle.

Multicore Support

If you come from the world of software development, you will be used to your programs being able to do more than one thing at a time – *multi-threading*, as it is called. This relies on the operating system switching between processes, to give the appearance of multiple things happening at the same time, or making use of multiple processors. Most microcontrollers don't use an operating system to schedule processes, and operate on a single processor (core). You can usually work around this limitation by using interrupts (as we saw earlier), but the ESP32 board can genuinely do two things at once.

The ESP32 is a dual-core (two-processor) device which means that, even without an operating system, it can assign activities to two different processes, allowing it to actually do two things at once. Often, in an embedded system, two things at once is about right – one eternal loop can be working the user interface (waiting for button presses and displaying things) while the other controls something.

We can rewrite our interrupt example to make use of two cores – one core just counting, while the other core waits for a button press. You will find this example in *12_04_multicore.py*.

```
from machine import Pin
from time import sleep
import _thread

switch = Pin(14, Pin.IN, Pin.PULL_UP)

def core0():
    x = 0
    while True:
        x += 1
        print(x)
        sleep(1)

def core1():
    while True:
        if switch.value() == 0:
            print("button pressed")
            sleep(0.1)

_thread.start_new_thread(core1, ( ))
core0()
```

A good way to separate what each core will do, is to define a function into which that core's code will go. So, in this case, core0 initialises a counter variable (x) to 0 and then loops forever, adding 1 to x and printing it out.

The other core (core1) has a second eternal loop, that watches for a switch press and prints out a message when it happens.

The function core1 is started by calling the start_new_thread method on _thread. This takes two parameters. The first is the name of the function to call, and the second must containing a tuple containing parameters to the function – here this is the odd-looking empty tuple ( ) as the parameter must be present even if you don't want to pass any parameters. `_thread` starts with an underscore as a way of signifying that the module is still experimental and not finalized.

`_thread` is just a way of identifying the second core (`core 1`).

The function `core0` is just started as a regular function call.

Note that this feature is flagged by the MicroPython community as *experimental*. I found it to work fine, but you may find that this may change. So if you have problems, then check with the latest MicroPython documentation here: `https://tinyurl.com/bderenhw`.

Analog Outputs

Although PWM is a pretty good mechanism for controlling the power to a load by varying the length of pulses, this carries with it the noise of the high frequency series of pulses. This doesn't matter a lot of the time, but an alternative feature of the ESP32 allows true analog outputs, where the voltage can be set to anything between 0V and 3.3V in 256 steps. The hardware on the ESP32 that does this is called a DAC (Digital to Analog Convertor) and is only available on pins 25 and 26.

If you have a digital multimeter, even a simple one, like the one shown in Figure 12.3, you can try this out. Set the multimeter to a voltage range greater than 3V (probably 20V), connect the negative lead to GND and the positive lead to pin 25. If your multimeter has alligator leads then clipping these to male-to-male jumper wires that can then connect to the breadboard is an easy way to do this.

Try running the program *12_05_dac_test.py*. This will prompt you to enter a voltage and, when you do so, the multimeter should show something close to the voltage entered, appearing on pin 25.

Figure 12.3. Measuring an analog output with a voltmeter.

```
from machine import DAC, Pin

dac = DAC(Pin(25))

while True:
    volt_str = input('Enter Voltage (0 to 3.3):')
    try:
        volt = float(volt_str)
        value = int(volt / 3.3 * 255.0)
        dac.write(value)
    except:
        print('bad voltage value')
```

Generating Sounds using the DAC

Since you can control the voltage at a DAC pin, you can use it to produce sounds from your ESP32 – but, you will need to connect a loudspeaker. The outputs from a GPIO pin can only supply a few milliamps, so you will need some kind of amplifier to drive the speaker. Here are a few amplifier/speaker combos that you can buy.

- MonkMakes Amplified Speaker 2 (https://monkmakes.com/amp_

spkr_2)

- Adafruit Mono 2.5W Amplifier (`https://www.adafruit.com/product/2130`) – external speaker required

- SparkFun Mono Audio Amp Breakout (`https://www.sparkfun.com/products/11044`) – external speaker required

Figure 12.4 shows the MonkMakes Amplified Speaker 2 attached to an ESP32 board. This board has the advantage that it has both amplifier and a small speaker attached to the same board.

Whatever amplifier you use the connections to be made are:

- GND on the ESP32 to GND on the Amplifier

- A power connection to 3.3V or 5V depending on the amplifier

- GPIO 25 to the audio input pin of the amplifier

Figure 12.4. Attaching an Amplified Speaker to an ESP32 Lite.

The program *12_07_play_sine_timer.py* will play a pleasing sine wave through the DAC on pin 25 (see Figure 12.5). When you run the program, you will be prompted to enter a frequency in Hertz (cycles per second). The number should be 50 to 500 – to produce an audible tone with a fairly accurate frequency.

Lets take a look at the code of *12_07_play_sine_timer.py*

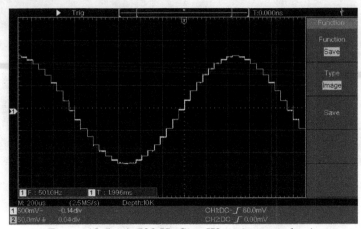

Figure 12.5. A 500 Hz Sine Wave (more or less).

```python
from machine import DAC, Pin, Timer
from math import sin, radians

dac = DAC(Pin(25))

sine_table = []
num_samples = 36
angle_per_sample = 360 / num_samples
scale_factor = 100
offset = 127

def fill_sine_table():
    global sine_table
    for i in range(0, num_samples):
        angle_degrees = i * angle_per_sample
        sine_table.append(int(sin(radians(angle_degrees)) *
            scale_factor + offset))

i = 0

def tick(timer):
    global i
    dac.write(sine_table[i])
    i += 1
    if i == num_samples:
        i = 0

fill_sine_table()
timer = Timer(0)

while True:
    f_str = input('Enter Frequency (Hz):')
```

```
try:
    f = int(f_str)
    sample_f = f * num_samples
    timer.deinit()
    timer.init(freq=sample_f, mode=Timer.PERIODIC, callback=tick)
except:
    print('bad frequency value')
```

The program first creates a table of sine-wave data for one single cycle (fill_sine_table), and then uses a hardware timer to read each value in turn from the table and write it to the DAC on pin 25. The data in sine_table is centred around the middle DAC value of 127 and the sine value (which is between 0.0 and 1.0) for the angle is multiplied by scale_factor to determine the volume to ensure a sensible amplitude/volume.

The constant num_samples determines how many sample values will be used to represent a single cycle of the sine wave. In this code num_samples is set to 36 and, if you look carefully at Figure 12.5, you can see that there are 36 steps making one cycle of the sine wave. Increasing this value will reduce the size of the steps and make a smoother waveform, but will reduce the maximum frequency that it is possible to play.

The timer calls the function tick at a frequency of the frequency chosen by the user. The timer can't call tick again until tick has finished, so if the frequency is increased too much, then some ticks will be lost and the frequency will be less than expected.

Then tick fetches the next value from sine_table, cycling back to the start of sine_table when it reaches the end.

To change the frequency of the timer in response to a new frequency being entered in the Shell, you have to first stop the timer using timer.deinit() before starting it again.

Playing a Sound File

The ESP32 has very little storage for things like sounds files. However, you can play short samples of sound at fairly low quality, such as the sound sample of a bell is included in the book downloads in the file *school_bell.raw*. This file needs to be copied onto the ESP32's file system; to do this, select *Files* from the *View* menu (Figure 12.6). In the top of the *Files* area navigate to the audio file on your computer, then right click on it and select *Upload to /*. This will copy the file onto the ESP32's file system.

Here is the code for the program *12_08_play_file.py*.

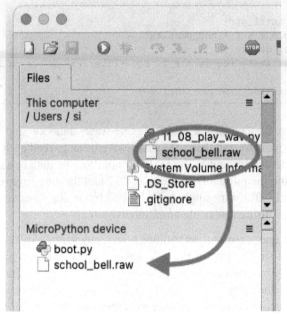

Figure 12.6. Copying the audio file onto the ESP32.

```
import time
from machine import DAC, Pin, Timer

filename = 'school_bell.raw'
dac = DAC(Pin(25))

buffer = []
num_samples = 0

def read_file():
    global buffer, num_samples
    try:
        f = open(filename, 'rb')
        buffer = f.read()
        num_samples = len(buffer)
        f.close()
    except:
        print("Couldn't find file: " + filename)

i = 0

def tick(timer):
    global i
    dac.write(buffer[i])
    i += 1
```

```
    if i == num_samples:
        timer.deinit()

read_file()

timer = Timer(0)
timer.init(freq=8000, mode=Timer.PERIODIC, callback=tick)

while True:
    pass
```

The code has much in common with the previous sine wave example but, in this case, the values to be written to the DAC successively are read from the file and the timer is set to call the `tick` function 8000 times a second for the audio sample rate of 8kHz.

Creating a Raw Sound File

The example program of *12_08_play_file.py* will only play an uncompressed mono 8kHz, 8-bit sound file.

If you want to create your own short sound file to play, then you can create a suitable file from a higher quality sample using the Audacity (`https://www.audacityteam.org/`) software. Once you have opened the audio file (in whatever format) in Audacity, you need to export is with some very specific settings, shown in Figure 12.7.

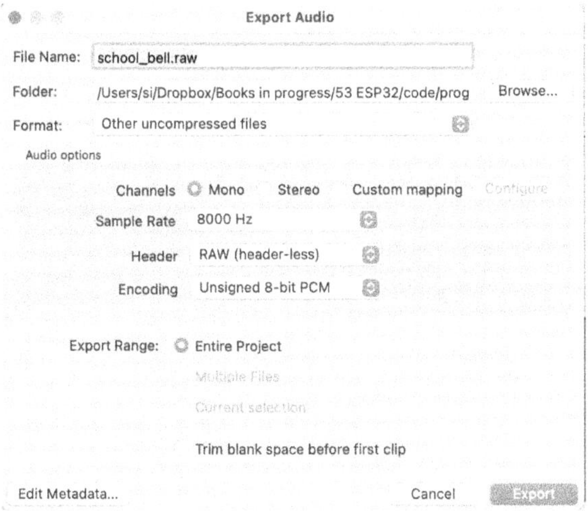

Figure 12.7. Audacity export settings.

From the *File* menu select *Export Audio...* In the *Format* drop-down select

Other uncompressed formats. Set *channels* to *mono*, *Sample Rate* to 8000, *Header* to *Raw (Headerless)* and finally *Encoding* to *Unsigned 8-bit PCM*.

Once you have the file in the correct format, you will have to save a copy onto your ESP32 as described earlier. If the file is too large, then you will see an exception when you try and run the program.

Touch

As we have seen, we can connect a push switch to a GPIO pin acting as a digital input; however, the ESP32 is also capable of detecting touch, opening up the possibility of making touch switches. Touch capability is available on the following pins: 0, 2, 4, 12, 13 14, 15, 27, 32 and 33.

Let's try this out on pin 32 using the code below, which can be found in the file *12_09_touch_test.py*:

```
from machine import TouchPad, Pin
from time import sleep

touch_pad = TouchPad(Pin(32))

while True:
    print(touch_pad.read())
    sleep(0.5)
```

When you run the program, you will see a stream of numbers in the Shell. Touch pin 32 with a finger and you should see that the number decreases. On my ESP32 Lite, I was seeing a no-touch reading of about 700 and a touch reading (touching a jumper wire connected to pin 32) of about 80 (although the existence of the jumper wire alone alters the reading too).

Unlike a digital input, when you define a TouchPad and read values from it using touch_pad.read(), you don't get a simple touched/not touched result. The number, you get, is like reading an analog input. The number you get depends on the size and shape of the touch pad connected to the GPIO pin. You could use a thumb-tack or a square of aluminum cooking foil connected to the pin with a wire. The touch sensing will also work if a reasonably-sized sensing area is covered by a thin layer of insulator, such as plastic. But you will need to know the threshold reading that should count as a touch press. You can use *12_09_touch_test.py*, to determine this value, and then put it in a program like *12_10_touch_led.py* which will turn on the built-in LED if touch on pin 32 is detected.

```
from machine import TouchPad, Pin, Signal
from time import sleep

touch_pad = TouchPad(Pin(32))
led = Signal(22, Pin.OUT, invert=True)

touch_thresold = 100

while True:
    if touch_pad.read() < touch_thresold:
        led.on()
    else:
        led.off()
```

GPIO Drive Current

The ESP32 has an interesting feature that allows you control the maximum drive current that a pin can supply. This also has the benefit that the pins are protected against accidental short-circuits by this current limiting device.

When defining a pin, you can supply an optional `drive` parameter like this, to set the maximum drive current:

```
p6 = Pin(32, Pin.OUT, drive=Pin.DRIVE_3)
```

In my experiments, the constants `Pin.DRIVE_0` to `Pin.DRIVE_3` set the current to between 11mA and 85mA. You can read more about this here: `https://tinyurl.com/4kz33bzz`.

Summary

In this chapter, we have discovered just how powerful the ESP32 board is when it comes to inputs and outputs. Not only can we do things pretty fast, but we can also use cores to run two programs at the same time.

The ESP32 is a fantastic little device, and is great for DIY maker projects. I hope this book has given you enough information to start your electronics and programming adventure with the ESP32 and MicroPython.

APPENDIX A

ESP32 Pinouts

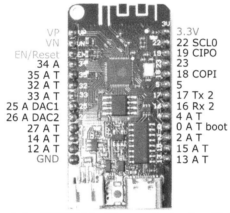

Figure A.1. ESP32 Lite Pinout. A=Analog, T=Touch

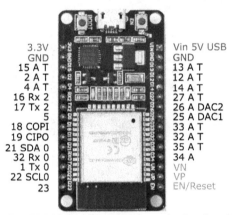

Figure A.2.   ESP32 DevKit 1 Pinout. A=Analog, T=Touch

APPENDIX B

Troubleshooting

Problem: I can't upload a program to my board and my board is not showing up as a USB device in the Thonny Preferences - Interpreter tab.

Solution: This usually means that you don't have the USB driver installed for your board's USB interface chip (see Page 7) or your board has developed a fault. Try restarting your computer and/or a different board.

I have found that sometimes on MacOS, boards will not show-up if connected directly using a USB C connector at the computer end. But, if a USB 2 hub is used instead, they work fine.

Problem: Thonny can connect to my board over USB, but I can't upload a new program, or get back to the Shell prompt.

Solution: Sometimes programs running on an ESP32 don't seem to want to quit. When you try and upload a new program, or interrupt the existing programe using the *Stop* button, it won't stop. Try setting the mouse focus to the Shell and press the CTRL and c keys simultaneously. Also try pressing the board's reset pin (or plug and unplug it) and press CTRL-c multiple times to try and catch a moment when the board is receptive. If all else fails, you can always reinstall MicroPython onto the board after putting it into download mode (see page 15).

Problem: My ESP32 Lite board has a built-in LED that blinks all the time.

Solution: This is the battery status LED and probably indicates that you don't have a battery attached to your board. If you don't have a need for a battery, then you will just have to ignore it; otherwise, see page 99 for use of a battery with the ESP32 Lite.

Problem: I'm trying to flash MicroPython onto my board using Thonny, but I keep getting errors.

Solution: Try reducing the *install speed* in the upload settings. Also, if the board shows up as two different interfaces in the *Target Port* list, try both options. Sometimes using a different USB port on your computer, or connecting via a USB 2 hub works.

Index

Biography

Simon Monk has written over twenty titles on programming and electronics and has sold over 750,000 books that have been translated into ten different languages. You can find out more about his books here: `http://simonmonk.org`.

Simon has a bachelor's degree in Computer Science and Cybernetics and a doctorate in Software Engineering. He has worked in many industries as a software developer and also pursues his interest in hobby electronics through the company MonkMakes Ltd (`https://monkmakes.com`).

He lives in the North West of England. You can follow him on X, where he is @simonmonk2